Optical Trapping and Manipulation of New Materials

Online at: https://doi.org/10.1088/978-0-7503-6074-6

Series Editor

R Barry Johnson, a Senior Research Professor at Alabama A&M University, has been involved for over 50 years in lens design, optical systems design, electro-optical systems engineering, and photonics. He has been a faculty member at three academic institutions engaged in optics education and research, has been employed by a number of companies, and has provided consulting services.

Dr Johnson is an IOP Fellow, an SPIE Fellow and Life Member, an OSA Fellow, and was the 1987 President of SPIE. He serves on the editorial board of *Infrared Physics & Technology* and *Advances in Optical Technologies*. Dr Johnson has been awarded many patents, has published numerous papers and several books and book chapters, and was awarded the 2012 OSA/SPIE Joseph W Goodman Book Writing Award for Lens Design Fundamentals (second edition). He is a perennial co-chair of the annual SPIE Current Developments in Lens Design and Optical Engineering Conference.

Foreword

Until the 1960s the field of optics was primarily concentrated in the classical areas of photography, cameras, binoculars, telescopes, spectrometers, colorimeters, radiometers, etc. In the late 1960s optics began to blossom with the advent of new types of infrared detector, liquid crystal display (LCDs), light emitting diode (LEDs), charge coupled device (CCDs), laser, holography, and fiber optics along with new optical materials, advances in optical and mechanical fabrication, new optical design programs, and many more technologies. With the development of the LED, LCD, CCD, and other electro-optical devices, the term 'photonics' came into vogue in the 1980s to describe the science of using light in the development of new technologies and the operation of a myriad of applications. Today optics and photonics are truly pervasive throughout society and new technologies are continuing to emerge. The objective of this series is to provide students, researchers, and those who enjoy self-education with a wide-ranging collection of books, each of which focuses on a topic relevant to the technologies and applications of optics and photonics. These books will provide knowledge to prepare the reader to be better able to participate in these exciting areas now and in the future. The title of this series is *Emerging Technologies in Optics and Photonics*, in which 'emerging' is taken to mean 'coming into existence', 'coming into maturity', and 'coming into prominence'. IOP Publishing and I hope that you will find this series of significant value to you and your career.

A list of recently published and forthcoming titles published in this series can be found here: https://iopscience.iop.org/bookListInfo/emerging-technologies-in-optics-and-photonics.

Optical Trapping and Manipulation of New Materials

Tiago de Assis Moura
Universidade Federal de Viçosa, Viçosa, Brazil

Joaquim Bonfim Santos Mendes
Universidade Federal de Viçosa, Viçosa, Brazil

Márcio Santos Rocha
Universidade Federal de Viçosa, Viçosa, Brazil

IOP Publishing, Bristol, UK

ISBN 978-0-7503-6074-6 (ebook)
ISBN 978-0-7503-6072-2 (print)
ISBN 978-0-7503-6075-3 (myPrint)
ISBN 978-0-7503-6073-9 (mobi)

DOI 10.1088/978-0-7503-6074-6

Version: 20250701

IOP ebooks

British Library Cataloguing-in-Publication Data: A catalogue record for this book is available from the British Library.

Published by IOP Publishing, wholly owned by The Institute of Physics, London

IOP Publishing, No.2 The Distillery, Glassfields, Avon Street, Bristol, BS2 0GR, UK

US Office: IOP Publishing, Inc., 190 North Independence Mall West, Suite 601, Philadelphia, PA 19106, USA

Contents

Preface

Although the optical tweezers technique is nowadays well known and used in many research laboratories around the world, today almost all applications still employ dielectric materials (usually microparticles) as handles to exert controllable forces on the systems of interest, allowing one to manipulate these systems and study their mechanical response. Such scenario is a consequence to the fact that dielectric materials can be more easily captured and stably trapped using highly focused laser beams—the basic optical tweezers setup.

In the past years, however, a series of works have emerged, with many groups interested in understanding the nature of the optical forces in different materials—from simple metals to semiconductors, topological insulators, and magnetic particles. In these cases, the stable trapping of particles made from such materials is not as straightforward as in the case of dielectric particles, and interesting particle dynamics occur as a consequence of the competition between different forces that arise from distinct physical phenomena.

In this book, we compile all the basic knowledge needed to understand the 'optical trapping and manipulation of new materials', also presenting some of the state-of-the-art research developed in the field along the past years. Our group of the Federal University of Viçosa (UFV, Brazil; www.lfb.ufv.br) is one of the pioneers in this field, starting to study the behavior of semiconductors and topological insulators in an optical trap in 2016 (being the first ones to report the oscillatory dynamics of such materials near the focal region of the laser beam).

In chapter 1, we present a brief field overview to introduce the research line that is the main subject of this book.

In chapter 2, we discuss the nature of the optical forces on dielectric materials in the Rayleigh regime (dipole approximation).

In chapter 3, we discuss geometrical optics approaches that are useful to calculate and predict optical forces on relatively large beads, considering important phenomena/issues such as light absorption by the beads and different laser intensity profiles.

In chapter 4, we discuss metallic materials and the field of plasmonics.

In chapter 5, we discuss the optics of semiconductors, which serve as an introduction to materials with intermediate properties between dielectrics and metals.

In chapter 6, we discuss the optical forces on semiconductors, presenting the relevant distinct phenomena involved and a model to calculate and predict these forces.

Finally, in chapter 7, we present and discuss some recent research concerning various different 'unusual' materials that present intermediate properties between dielectrics and conductors.

This book was prepared by the three authors giving different contributions. Márcio S Rocha wrote chapter 1, with the contribution of Tiago A Moura in the first section. Tiago A Moura wrote chapters 2, 4, 5, and 6. Márcio S Rocha wrote chapters 3 and 7 (this last one with the contribution of Joaquim B S Mendes). Joaquim B S Mendes wrote the appendix. All authors developed the authorial research reported in the book. All authors contributed to revise the entire manuscript.

Acknowledgments

The authors would like to express their sincere gratitude to all those who supported the development and completion of this project. They are especially thankful to their collaborators and colleagues whose critical insights and shared expertise significantly enriched the scope and rigor of this work. The exchange of ideas and the collaborative spirit within the scientific community have been invaluable throughout this endeavor.

They gratefully acknowledge the Universidade Federal de Viçosa (UFV), particularly the Department of Physics, as well as the Brazilian funding agencies CAPES, FINEP, CNPq, and FAPEMIG for providing the resources, funding, and infrastructure essential to this project's success.

They also extend their heartfelt appreciation to their families and friends for their patience, understanding, and unwavering support during the many hours devoted to research and writing.

Finally, they recognize the pioneering contributions of researchers in the fields of optical trapping and materials science, whose foundational work continues to inspire innovation and discovery.

Author biographies

Tiago de Assis Moura (T A Moura)

Departamento de Física, Universidade Federal de Viçosa, Viçosa, Minas Gerais, Brazil.

Graduated (2015), Master's (2017), and PhD (2022) in physics from the Federal University of Viçosa (UFV), Brazil. Since 2017 he has been studying the interaction and manipulation of semiconductor materials with the optical tweezers technique. Recently, he was a postdoctoral fellow of the Center for Research in Energy and Materials (CNPEM, Brazil), working at the Federal University of Viçosa under the supervision of J B S Mendes and M S Rocha.

Joaquim Bonfim Santos Mendes (J B S Mendes)

Departamento de Física, Universidade Federal de Viçosa, Viçosa, Minas Gerais, Brazil.

Joaquim Mendes holds a master's degree in Physics from the Federal University of Pernambuco (UFPE, 2009) and a PhD in Physics from the Federal University of Rio de Janeiro (UFRJ, 2013). He was a post-doc in the Physics Department of UFPE (2014). Since 2014 he has been a professor (permanent position) at the Physics Department at the Federal University of Viçosa (UFV) in Minas Gerais, Brazil. He was director of the Physics Department at UFV from 2019 to 2025. He has experience in experimental physics, with emphasis on condensed matter physics, materials science, magnetic materials, and magnetic properties, working mainly on the following topics: nanomagnetism and spintronics, ferromagnetic resonance, spin dynamics, quantum materials, optical manipulation of micro-systems, synthesis of nano and microparticles, fabrication of thin films and multilayers, etc.

Márcio Santos Rocha (M S Rocha)

Departamento de Física, Universidade Federal de Viçosa, Viçosa, Minas Gerais, Brazil.

Graduated in Physics from the Federal University of Ceará (2001), Master's (2004), and PhD (2008) in physics from the Federal University of Minas Gerais, both Brazilian institutions. Since 2008 he has been a professor (permanent position) at the Physics Department at the Federal University of Viçosa (UFV) in Minas Gerais, Brazil, becoming a Full Professor in December 2024. He was coordinator of the Graduate Program in Physics at UFV from 2017 to 2022.

He has extensive experience with experimental physics, with emphasis on the fields of biological physics, optics applied to biological systems, and optical manipulation of micro-systems. He is Head of the Biological Physics Laboratory at UFV, implemented by the researcher at this institution in 2008–10 and Director of the Nanoscopy Research Center of UFV (2024–present). Lab website: http://www.lfb.ufv.br.

IOP Publishing

Tiago de Assis Moura, Joaquim Bonfim Santos Mendes and Márcio Santos Rocha

Chapter 1

Field overview

In this chapter, we present an overview of the field that is the subject of this book: the optical trapping and manipulation of materials. In particular, we discuss some key features concerning the trapping of dielectric particles, which are still the main handles used in most optical tweezers (OT) experiments. A helpful discussion on the effects of light absorption by these particles is presented, a feature that is many times neglected. Furthermore, the main approaches used to calculate and predict the resultant optical forces on these particles are discussed, along with some experimental details for performing accurate force measurements with OT.

1.1 A brief history of OT

Johannes Kepler (1571–1630) was one of the first scientists to intuit that light would be able to exert forces on objects. He believed that the Sun's rays, acting on the tail of comets, would sweep them away in a direction always away from the Sun.

A few centuries later, the Scottish physicist and mathematician James Clerk Maxwell (1831–79), through his brilliant work in electrodynamics, demonstrated that Kepler was correct. However, the forces exerted by light beams are usually too insignificant to be observed acting on macroscopic objects.

With the emergence of the laser in 1960, capable of concentrating a large amount of light into a single beam, the possibility of using light as a tool for manipulating objects on a microscopic scale began to emerge. It was Arthur Ashkin (1922–2020) who first utilized lasers to manipulate and trap small dielectric particles in 1969 while working at the Bell Laboratories in the United States, ushering in a revolution in both physics and biology [1]. In particular, attention was drawn to the significant acceleration that particles acquired when colliding with a laser beam, even with low-power lasers (the concentration of the rays ensures effectiveness).

In the beginning there was a great interest in using light to manipulate and trap atoms. However, it was from 1987 that OT found some of their greatest utilities in the field of biology [2, 3]. In simple terms, OT can be understood as highly focused

doi:10.1088/978-0-7503-6074-6ch1

laser beams, usually obtained using the objective lens of a microscope. Near the focal region of the objective, particles, typically dielectric in nature, can be trapped and manipulated, serving as tools in various applications. Among the numerous applications of OT in biology, noteworthy examples include the trapping of viruses and bacteria [2], manipulation of individual cells [4], investigation of mechanical properties of cell membranes [5], investigation of proteins [6], analysis of DNA–ligand interactions [7], among others [8].

The following sections provide insightful explanations on how optical forces arise and enable the trapping of particles in various optical regimes. Light can transfer momentum and energy during its interaction with matter. Optical forces, in general, can be categorized into two types: those arising from momentum transfer and those resulting from energy transfer, often occurring indirectly.

The so-called OT [9] refer to traps created using highly focused laser beams. They utilize the momentum transfer between the laser and semi-transparent particles, typically dielectrics, to trap them close to the focal region. The technique works very well for dielectric particles using ordinary Gaussian (TEM_{00}) beams provided that the focusing is sufficiently strong, which can be done in practice by using high numerical aperture microscope objectives. For other types of particles, however, light scattering and/or absorption can be a real problem.

To minimize scattering effects, specific beam configurations are usually employed, such as those with radial polarization [10] or dark centers [11][1]. These techniques can be applied to trap metallic particles in both the Rayleigh and in the geometrical optics regime [12], which will be defined soon.

For particles that considerably absorb light when immersed in a fluid (liquid or gas), the use of momentum transfer to trap them can be significantly impaired. This is because absorption usually leads to the emergence of a fluid dynamics force acting on the particles. This force, which results from the asymmetric energy transfer from light to the particle, is known as the photophoretic force [13] and will be discussed in section 1.5.

Until the late 1970s, optical traps were primarily constructed using a configuration involving two or more beams positioned opposite each other. These traps relied on the radiation pressure exerted by the beams for trapping [14]. However, in 1978 [15], while attempting to trap and cool atoms, Ashkin discovered the existence of a 'net reverse radiation pressure force' in strongly focused beams[2]. This breakthrough opened the door for the development of single-beam traps.

In a seminal work in 1986 [16], Ashkin demonstrated the effectiveness of single-beam traps for particles that were significantly smaller in size compared to the laser wavelength. Furthermore, in 1992 [17], Ashkin demonstrated and explained the trapping of particles with dimensions much larger than the wavelength of the laser

[1] Dark center beams are characterized by having zero-intensity along the optical axis. Examples are Bessel beams of higher orders.

[2] Nowadays this force is commonly referred to as the gradient force, as it originates from the gradient of electric field intensity.

utilizing a simple ray optics model. These advancements marked important milestones in the field of optical trapping and manipulation.

From 1986 onwards, following the successful application of OT in biology, several researchers have revisited Ashkin's initial work and proposed corrections and enhancements to adapt the theory to experimental conditions. In this way, the works by Barton *et al* in 1988 [18] and 1989 [19] proposed improvements to the paraxial approximation[3] for strongly focused Gaussian beams. These corrections enabled a more comprehensive description of the incident and scattered fields, based on the vector wave equation and a previous work by Davis in 1979 [20]. In parallel, Gouesbet *et al* published works in 1988 [21], 1994 [22], and 1996 [23] in which they generalized the Lorenz–Mie scattering theory to describe the interaction of particles of arbitrary sizes with beams of arbitrary shapes, including highly convergent Gaussian beams commonly used in OT. Finally, another theory of OT developed by Maia Neto and Nussenzveig appeared in 2000 [24] based on expanding the field in partial waves (Mie) to describe the force exerted on a transparent sphere by a Gaussian laser beam focused through a high numerical aperture objective. This model successfully covered the entire range of interest for OT applications, predicting behaviors in both the Rayleigh and geometric optical limits. It offered the potential for absolute calibration of OT and was highly improved over the years by the group, including corrections and optical aberrations. These theories of OTs will be further discussed later in this chapter.

1.2 Dielectric beads: the standard tools for OT

In past years, micrometer-sized semi-transparent dielectric spheres, usually referred to as microspheres or beads, have become the standard tools in most OT applications. Although the main motivation for using this type of beads was already mentioned in the former section, let us formally cite the main reasons for that:

- Microspheres (beads) present the perfect symmetrical shape, which highly facilitates the theoretical calculation of the optical forces acting on them for various different situations concerning distinct bead size ranges, beam types, etc. In addition, from the experimental point of view, the optical forces that a tweezer exerts on such beads can be measured with high accuracy.
- Semi-transparent dielectric beads made of common materials such as polystyrene or silica usually refract light very well, presenting small absorption coefficients. The reflection of light on the surface of such beads, on the other hand, can also be made small using appropriate beam geometries. Thus, by using these beads one can highly improve the attractive gradient forces relatively to repulsive scattering and photophoretic forces, facilitating the three-dimensional (3D) trapping and manipulation of such beads (as will be evident throughout this book).

[3] The paraxial approximation is based on the assumption that when a beam propagates over a distance equal to the wavelength, the variation in intensity is negligible compared to the overall intensity of the field. In the context of geometrical optics, this implies that the rays that can be drawn to represent the beam are practically parallel along the beam propagation.

- The types of beads above described, (i.e. polystyrene or silica beads) can be bought from many manufacturers in monodisperse solutions having highly accurate calibrated sizes, with diameters that can be chosen from nanometers to tens of micrometers. Furthermore, they can also be bought with their surfaces functionalized, i.e. coated with special molecules like proteins or other reactive groups that can be attached on biological or other types of systems that one intends to study such as biopolymers, biological membranes and surfaces, among others.

In the next section, we discuss briefly the main theoretical approaches that can be used to calculate the optical forces exerted by tweezers on dielectric beads. Such simple models serve as the basis for more elaborate theories and can give helpful (and many times accurate) insights on the behavior of such beads in an optical trap.

1.3 Optical forces on dielectric beads

Over the years in fact, many different theoretical approaches were proposed to calculate and predict the optical forces exerted by a focused laser beam on dielectric beads. Most of these theories neglect light absorption and consider the optical forces associated with light reflection and refraction on the bead. For didactic purposes and to give the reader a general overview on the theory of optical forces, we discuss briefly the main approaches usually employed to calculate the optical forces on dielectric particles.

1.3.1 Rayleigh regime

As the name suggests, this regime in practice is valid for very small particles, with typical sizes much smaller than the laser wavelength. We will discuss such approximation in greater detail in chapter 2. For now, it is sufficient to state the simplified mathematical condition that should be satisfied for the accuracy of the present approach [24, 25]

$$k_{\mathrm{m}} a \ll 1, \tag{1.1}$$

where k_{m} is the wave number in the medium surrounding the bead, which has a radius a. Observe that this condition above implies that, for most practical purposes, the bead radius must be much smaller than the wavelength of the laser used in the tweezers, as commented before.

The trapped bead being very small, it can be approximately treated as an induced electric dipole in the presence of the laser radiation. In a first-order approximation, we can consider that the spatial variation of the electric field of the beam around the bead volume is very small and thus consider this field as spatially uniform around the bead [9], which considerably simplifies the calculations. In fact, problems concerning the presence of dielectric spheres inside uniform electric fields are basic tasks found in books on classical electrodynamics [26]. Straightforward calculations show that the induced electric dipole moment on such a dielectric sphere can be written as

$$\mathbf{p} = 4\pi\varepsilon_0\varepsilon_m\frac{\varepsilon_r - 1}{\varepsilon_r + 2}a^3\mathbf{E}, \tag{1.2}$$

where $\mathbf{E}$ is the original electric field in the absence of the bead, ε_m is the dielectric constant of the surrounding medium (supposed to be linear and isotropic), ε_0 is the electric permitivitty of the free space, and ε_r is the relative dielectric constant, given by

$$\varepsilon_r = \frac{\varepsilon_p}{\varepsilon_m}, \tag{1.3}$$

where ε_p is the dielectric constant of the particle.

With equation (1.2) we can find the polarizability of the bead α, which is a constant of proportionality between the induced dipole moment and the electric field [26],

$$\mathbf{p} = \varepsilon_0\varepsilon_m\alpha\mathbf{E}. \tag{1.4}$$

By comparing equations (1.4) and (1.2), one can thus promptly write

$$\alpha = 4\pi\frac{\varepsilon_r - 1}{\varepsilon_r + 2}a^3 = 3V\frac{\varepsilon_r - 1}{\varepsilon_r + 2}, \tag{1.5}$$

where V is the volume of the bead. This equation is known as the *Clausius–Mossotti relation.*

Electromagnetic theory also predicts the force on a dipole in the presence of an electric field to be [26]

$$\mathbf{F} = (\mathbf{p} \cdot \nabla)\mathbf{E}. \tag{1.6}$$

In the case of an electrostatic field, this equation can be rewritten in a more convenient form using the vector identity

$$\nabla(\mathbf{A} \cdot \mathbf{B}) = \mathbf{A} \times (\nabla \times \mathbf{B}) + \mathbf{B} \times (\nabla \times \mathbf{A}) + (\mathbf{A} \cdot \nabla)\mathbf{B} + (\mathbf{B} \cdot \nabla)\mathbf{A}. \tag{1.7}$$

Making $\mathbf{A} = \mathbf{p}$ and $\mathbf{B} = \mathbf{E}$ and remembering that $\nabla \times \mathbf{E} = 0$ in electrostatics, one can promptly write[4]

$$\mathbf{F} = \frac{1}{2}\varepsilon_0\varepsilon_m\alpha\ \nabla\ (\mathbf{E} \cdot \mathbf{E}). \tag{1.8}$$

Therefore, using equations (1.2) and (1.8), one finds

$$\mathbf{F} = 2\pi\varepsilon_0\varepsilon_m\frac{\varepsilon_r - 1}{\varepsilon_r + 2}a^3\ \nabla\ E^2. \tag{1.9}$$

It should be noted that this force is proportional to the gradient of the field intensity ($\propto E^2$), meaning that it points to the region of maximum intensity of the field. Therefore, for a rigorously uniform field no net force would actually be exerted on the bead. On the other hand, if one creates a small region of maximum intensity

---

[4] In fact, this condition is more general because, according to Faraday's law, $\nabla \times \mathbf{E} \propto -\partial_t\mathbf{H}$, and when dealing with optical fields, where the oscillation frequency is high, only the time-averaged fields are physically meaningful, leading to $\nabla \times \mathbf{E} = 0$. We will discuss the details in chapter 2.

by focusing the laser beam, equation (1.9) suggests that the beads should be attracted to that region if $\varepsilon_r > 1$ (or in other words, if the bead has a refractive index higher than that of the medium). This is actually what is experimentally observed when using dielectric Rayleigh particles in a single-beam OT, which consists of a laser beam highly focused by a lens.

Due to the behavior of this type of force, which is proportional to the gradient of the field intensity, it is usually referred to as a *gradient force*. Note that gradient forces act as a restoring force, bringing the particle back to an equilibrium position when it is displaced from that position. For small bead displacements (first-order approximation) one expects that restoring forces should be proportional to the displacement from the equilibrium position, such that a bead trapped by a focused laser beam behaves as a harmonic oscillator due to the action of the restoring gradient forces. Gradient forces exerted by a focused laser beam on a bead are conservative forces (remember that we neglected the curl of the electromagnetic field in the derivation of equation (1.8)), as one should expect due to the similarity with a simple spring–mass system.

It is usually important to characterize the spring constant associated to the harmonic oscillator above discussed (laser–bead system), which we call the *trap stiffness* of the OT. For a bead displacement in a given direction (say, the Cartesian coordinate x), for example, the x-component of the force will be

$$F_x = 4\pi\varepsilon_0\varepsilon_m \frac{\varepsilon_r - 1}{\varepsilon_r + 2} a^3 \frac{\partial E^2}{\partial x}, \tag{1.10}$$

and the trap stiffness can be written as

$$\kappa_x = \left(\frac{\partial F_x}{\partial x}\right)_{\mathbf{r}_{eq}} = 4\pi\varepsilon_0\varepsilon_m \frac{\varepsilon_r - 1}{\varepsilon_r + 2} a^3 \left(\frac{\partial^2 E^2}{\partial x^2}\right)_{\mathbf{r}_{eq}}, \tag{1.11}$$

where $\mathbf{r}_{eq}$ is the equilibrium position of the trapped bead.

Observe that in the Rayleigh regime, the trap stiffness is proportional to the cubic power of the bead radius, i.e. it increases proportionally to a^3 in this regime.

Despite the approximations done above when deriving the gradient force in the Rayleigh regime, equation (1.9) works surprisingly well. To improve the accuracy of the model, one can simply exchange the static polarizability (equation (1.5)) by the real part of the complex dynamic polarizability, since in the real situation the electric field of the incident light is oscillating with a frequency ω.

A derivation of the complex dynamic polarizability can be found in references [27, 28] and the result is

$$\alpha_{dyn} = \alpha\left[1 - \frac{\varepsilon_r - 1}{\varepsilon_r + 2}\left(k_0^2 a^2 + \frac{2i}{3} k_0^3 a^3\right)\right]^{-1}, \tag{1.12}$$

where $k_0 = 2\pi/\lambda_0$ is the wavenumber in the vacuum.

Furthermore, the electric field of an electromagnetic wave does not present a null curl and therefore other important terms in equation (1.8) will appear when suppressing the quasi-static approximation for the electric field. We will not reproduce such calculations here, but an appropriate discussion will be done in chapter 2. The two additional terms are the *scattering force* and the *spin-curl force*.

The spin-curl force arises from spatial changes of the polarization in the electromagnetic field and is typically much smaller than the gradient and scattering forces [27]. Therefore, it can be safely neglected in most OT experiments.

Scattering forces, on the other hand, present a fundamental importance in OT experiments; not only in the Rayleigh approximation, but also in other regimes of particle size. For small Rayleigh particles, the calculations mentioned above show that this force can be written as [27]

$$\mathbf{F}_S = \frac{\sigma_{\mathrm{ext}}}{c}\langle \mathbf{S} \rangle, \tag{1.13}$$

where σ_{ext} is the extinction cross-section, c is the light speed, and $\langle \mathbf{S} \rangle$ is the time-averaged Poynting vector of the incoming electromagnetic field.

Note that scattering forces point in the direction of light propagation and thus tend to push the particle along this direction. Furthermore, they are also proportional to the extinction cross-section, which shows that it is related to light reflection and absorption on the bead. Therefore, such forces tend to disturb the trapping and, in practical situations, need to be smaller than the gradient forces in order for one to achieve stable particle trapping. This is why most OT experiments are usually performed using semi-transparent dielectric beads, which present small reflection and absorption coefficients, largely refracting the incident light. As will be evident in chapter 3, light refraction by the bead is closely related to the generation of gradient forces, which are the key type of force for stable trapping.

1.3.2 Geometrical optics (GO) regime

This approach corresponds to the opposite asymptotic limit considered in the Rayleigh regime: here we are interested in beads much larger than the laser wavelength, such that GO (also called ray optics) calculations can be applied to compute the beam interaction with the object and consequently to predict the optical forces. Formally, the mathematical condition for this approach to hold can be written for practical purposes as

$$k_{\mathrm{m}} a \gg 1, \tag{1.14}$$

where the variables were previously defined.

Geometrical optics calculations can be done basically by computing the momentum change of a single ray of light reaching the object due to its reflection, refraction, and absorption, and then summing (integrating) over the entire laser beam [9, 17, 29, 30]. Effects such as some optical aberrations, specially spherical aberration, can also be taken into account in such calculations [9], returning more accurate results. In chapter 3, we will discuss GO calculations in detail, presenting the basic formalism that can be used to compute the optical forces on the beads. Such calculations allow one to associate gradient forces with light refraction on the bead and scattering forces with light reflection and/or absorption by the bead.

For now let us cite the main result: GO calculations predict that the trap stiffness of the OT exhibits a hyperbolic decrease as a function of the bead radius, i.e. it is proportional to $1/a$ [9]; a behavior completely different from that found for the Rayleigh regime discussed above in which the trap stiffness increases proportionally to a^3.

Although limited to larger particle sizes, GO calculations are versatile and easy to be implemented numerically, allowing one to study how the optical forces, and consequently the trap stiffness vary as a function of important parameters such as bead radius, refractive index, and absorption coefficient, for example. Such calculations can also be implemented for different beam geometries, including distinct transverse intensity profiles and optical aberrations, if present in the system [9, 31].

1.3.3 General approaches

The two asymptotic approaches described above (Rayleigh and GO) are useful when dealing with small ($a \ll \lambda$) or large ($a \gg \lambda$) beads, respectively, but both can fail in the intermediate region when $a \sim \lambda$. In past years, authors have proposed more general approaches, which intend to account for all particle size ranges in relation to the laser wavelength used in the OT setup.

The force determination in all of these models is based in calculating the Maxwell stress tensor, a quantity used to describe the interaction between the electromagnetic field and a material object inside this field. Such a tensor appears naturally when considering the conservation laws involving the linear and the angular momentum of the system 'light + object', as can be found in many textbooks on electrodynamics [26]. It can be written as

$$\overleftrightarrow{T_{\mathrm{M}}} = \varepsilon_0\left(\mathbf{E} \otimes \mathbf{E} - \frac{1}{2}E^2\overleftrightarrow{I}\right) + \frac{1}{\mu_0}\left(\mathbf{B} \otimes \mathbf{B} - \frac{1}{2}B^2\overleftrightarrow{I}\right), \tag{1.15}$$

where $\mathbf{E}$ and $\mathbf{B}$ are the electric and magnetic fields, ε_0 and μ_0 are the electric permittivity and the magnetic permeability of the medium (free space in the case), $\otimes$ denotes the dyadic product which results in a tensor represented by a 3×3 matrix and $\overleftrightarrow{I}$ is the identity tensor (3×3 identity matrix).

In terms of its components, the Maxwell stress tensor can also be written as

$$T_{\mathrm{M}}^{ij} = \varepsilon_0\left(E_iE_j - \frac{1}{2}E^2\delta_{ij}\right) + \frac{1}{\mu_0}\left(B_iB_j - \frac{1}{2}B^2\delta_{ij}\right), \tag{1.16}$$

where $i, j = x, y, z$ in Cartesian coordinates and δ_{ij} is the Kronecker delta.

With this tensor, the radiation force $\mathbf{F}_{\mathrm{rad}}$ on an object (e.g. a bead in the field) can be calculated as an integral over a surface that surrounds the object

$$\mathbf{F}_{\mathrm{rad}} = \oint \langle \overleftrightarrow{T_{\mathrm{M}}} \rangle \cdot \hat{n} dA, \tag{1.17}$$

where $\langle \overleftrightarrow{T_M} \rangle$ denotes the time-averaged Maxwell stress tensor, $\hat{n}$ is an unit vector normal to the surface of integration and pointing outwards and dA is the area element.

Furthermore, the Maxwell stress tensor can also be used to calculate the radiation torque on the object Γ_{rad}, which is given by [27]

$$\Gamma_{rad} = -\oint (\langle \overleftrightarrow{T_M} \rangle \times \mathbf{r}) \cdot \hat{n} dA. \tag{1.18}$$

To apply equations (1.17) and (1.18) in the calculation of the force and torque on a particle inside an electromagnetic field, one needs to account for the total electric and magnetic fields on the surface of integration, which consist in the sum of two components each: the incident and scattered fields by the particle. Therefore, in practice, one needs to solve the associated problem concerning light scattering by the particle. Furthermore, in the specific case of OT, one also needs to accurately write expressions for the strongly focused incident fields.

In this section, we briefly comment on some of the most known approaches used to calculate the radiation forces by directly applying the Maxwell stress tensor. The formal development of the calculations can be found in the original articles, properly cited in the discussion.

1.3.3.1 Transition matrix (T-matrix) method

In this approach, the components of the incident ($\mathbf{E}_i$) and scattered ($\mathbf{E}_s$) fields are related by an operator, the *transition matrix* $\mathscr{T}$,

$$\mathbf{E}_s = \mathscr{T}\mathbf{E}_i. \tag{1.19}$$

The fields $\mathbf{E}_i$ and $\mathbf{E}_s$ can be expanded in terms of multipoles, such that one can deduce a transition matrix $\mathscr{T}$ that relates the incident and scattered field components, which carry all the information about the particle shape and orientation in relation to the incident field [25, 27]. This method is computationally effective because it allows the use of the rotation and translation properties of the T-matrix to evaluate the optical forces and torques for various configurations (distinct positions and orientations of the trapped object) [25]. Finally, to represent the incident field, the incoming beam is expanded into plane waves and its focusing is written in the angular spectrum representation [25].

The T-matrix calculations are more accurate when dealing with perfectly symmetric particles, which should also present a homogeneous composition. However, it is possible to calculate forces and torques on asymmetrical particles by modeling them as aggregates of smaller spheres [25].

1.3.3.2 Generalized Lorenz–Mie theory (GLMT)

Another way to express the fields as multipole expansions is provided by the GLMT [32], in which an arbitrarily shaped beam is expanded in vector spherical harmonics, such that the scattering of this field could be determined for a symmetric object (e.g. a spherical bead). The optical forces can then be expressed in terms of the beam shape coefficients (BSCs), which carry information about the structure of the incident laser beam.

This type of approach has been developed by Gouesbet and collaborators since the end of the 1980s, and various generalizations and improvements were proposed in subsequent years [32]. It is worth saying that the two discussed methods (T-matrix and GLMT) are not independent, and in fact can be connected [33].

1.3.3.3 Mie–Debye spherical aberration (MDSA) model and generalizations
This approach is based on the following main statements:

- The model is formulated using a Debye-type integral representation for the focused laser beam as a superposition of plane waves [3], first developed by Richards and Wolf [34]. This representation presents the great advantage of correctly describing a highly focused beam such as those typically found in OT setups (due to the high numerical aperture objectives that are usually employed). Furthermore, the representation also takes into account the beam diffraction at the back aperture of the objective lens.
- The field scattered by the particle is calculated using the Mie scattering theory and thus is valid for beads of any size and refractive indices.
- The model correctly accounts for the objective transmittance, allowing one to calculate the local laser power on the bead by knowing the input laser power at the back aperture of the objective [35].
- The MDSA model does not use any adjustable parameter: all the parameters needed to calculate the trap stiffness for a given experimental setup can be measured independently and used in the calculations as input parameters [36].
- The calculations can consider the main relevant optical aberrations present in most OT setups, in special spherical aberration and astigmatism. In the past years, the authors have demonstrated that the considerations of such aberrations is crucial for one to get accurate theoretical results, which reproduce the measured values of the trap stiffness for various experimental conditions [36–38].

In figure 1.1, we show an absolute (no adjustable parameters) comparison between the theoretical prediction of the MDSA model and experimental results for a specific OT setup described in the original reference [38]. The data shows the transverse trap stiffness κ_x normalized by the laser power P as a function of the bead radius a. The *red points* are experimental data with the corresponding error bars. The *dashed line* is the prediction of the original MDSA theory considering spherical aberration as the only optical aberration present in the setup. The solid line is the prediction of the MDSA+ model, an extension of the original MDSA which also includes the astigmatism aberration, for which the characterizing parameters were determined experimentally and used in the theoretical calculations [38]. Observe that the agreement between the experimental data and the model is excellent when one takes into account the relevant optical aberrations present in the setup.

To conclude this section, we stress that the models presented and the discussions performed show that a complete and accurate OT theory must take into account

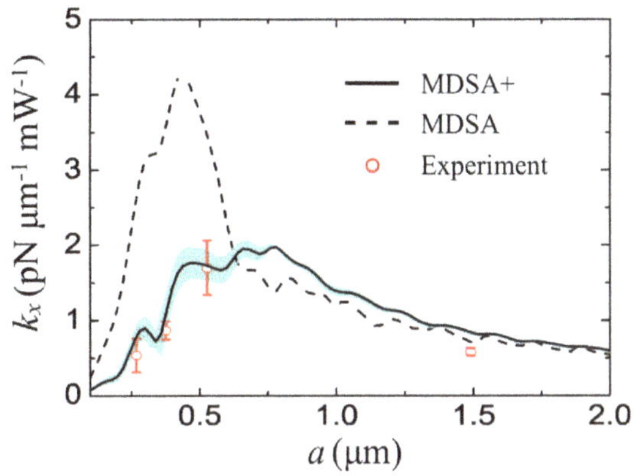

Figure 1.1. Absolute (no adjustable parameters) comparison between the theoretical prediction of the MDSA model and experimental results for a specific OT setup [38]. The data shows the transverse trap stiffness κ_x normalized by the laser power as a function of the bead radius a. The *red points* are experimental data with the corresponding error bars. The *dashed line* is the prediction of the original MDSA theory considering spherical aberration as the only optical aberration present in the setup. The solid line is the prediction of the MDSA+ model, an extension of the original MDSA which also includes the astigmatism aberration. Reprinted from [38], with permission of AIP Publishing.

various features, starting from a robust characterization of the focused beam up to the inclusion of relevant optical aberrations that should be characterized for each setup, allowing the inclusion of the characterizing parameters in the theoretical calculations. One should note that there is no need to use adjustable parameters in an OT theory: all parameters needed to calculate the trap stiffness for a given setup can be measured independently and used in the calculations.

1.4 Performing measurements with dielectric beads

Parts of this section have been reproduced from [45].

From the discussions performed in the former sections, it is evident that there are two relevant optical forces that one deals with when using semi-transparent dielectric beads in OT: scattering forces, which arise from light reflection (and absorption, which is usually small for these materials) by the bead and gradient forces, which arise from light refraction on the bead, as will be evident in chapter 3. Scattering forces tend to push the objects along the direction of the laser propagation, disturbing the stable trapping. In fact, if the intention is to stably trap an object at the focal region, gradient forces must supplant scattering forces. Therefore, for being stably trapped at the focus, it is desirable that the bead presents small reflection and absorption coefficients, and consequently a high transmission coefficient, which in practice occurs for semi-transparent dielectric objects, as anticipated. Metallic conducting beads, on the other hand, usually present high reflection and absorption coefficients, which makes stable trapping difficult. Furthermore, absorbing particles are also subjected to photophoretic forces, which arise from heating effects and tend to highly disturb the trapping, as will be discussed in section 1.5. Finally, it is worth commenting that for particles made of materials with intermediate properties (e.g. semiconductors), the resulting behavior will depend on a delicate balance between these two optical forces and the photophoretic forces, as will be evident in the following chapters.

Gradient forces tend to attract dielectric particles to the focal region regardless of their initial position in the beam, provided that they present a refractive index higher than that of the surrounding medium. When stably trapped at the focus, if an external perturbation acts on the bead changing its position, the gradient force will act as a restoring force bringing it back to its equilibrium position. For a small bead displacement along a given axis (say, x) relative to the focus, the restoring optical force can be written as

$$F = -\kappa \Delta x, \tag{1.20}$$

where Δx is the distance between the bead center and the focus, and κ is the trap stiffness previously defined.

Most OT experiments are performed with the bead embedded in a solvent (usually water-based solutions) such that, besides the optical forces already discussed, the random forces associated with its Brownian motion play an important role in the bead dynamics, as well as the friction forces between the bead and the solvent. Therefore, a microscopic particle trapped by OT inside a solution is typically an *overdamped Brownian harmonic oscillator*, characterized by a parabolic potential well given by $U(x) = \kappa x^2/2$ (valid for small displacements). Figure 1.2

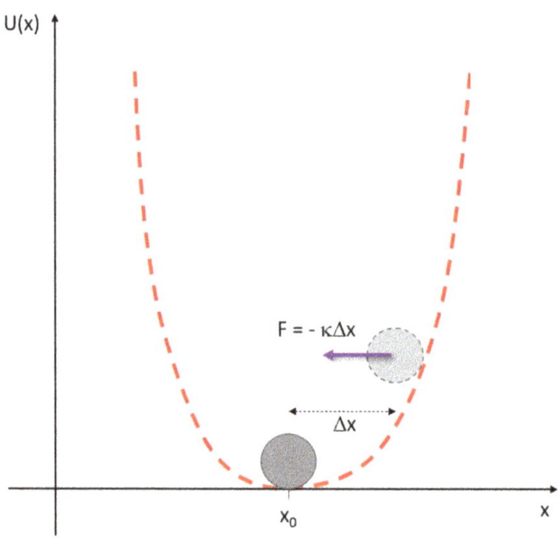

Figure 1.2. A microscopic particle trapped by an OT inside a solution is typically an *overdamped Brownian harmonic oscillator*, characterized by a parabolic potential well given by $U(x) = \kappa x^2/2$ (associated with the conservative gradient force). Here, x_0 is the equilibrium position of the bead at the bottom of the potential. A net displacement relative to this equilibrium position promoted by an external agent can then be expressed by $\Delta x = x - x_0$. The restoring force that brings the bead back to its equilibrium position is given be equation (1.20). Adapted from [45], with permission from Springer Nature.

represents a bead inside this parabolic potential well, which is associated with the gradient force—the only type of optical force that is conservative. We call x_0 the equilibrium position of the bead at the bottom of the potential. A net displacement relative to this equilibrium position promoted by an external agent can then be expressed by $\Delta x = x - x_0$.

Force calculation and measurements are highly facilitated if the particles present perfect spherical shapes. In fact, optical forces on spherical shaped beads can be measured in the laboratory with a high accuracy using equation (1.20) if one knows the trap stiffness κ associated with the experimental setup. Note that one also needs to monitor and measure the bead position with accuracy, which can be done using a position detector or videomicroscopy, a technique that consists in using a camera to record the experiments for posterior digital image analyses.

The procedure of measuring the trap stiffness κ of an OT is usually referred to in the literature as the *calibration of the tweezers* and must be performed before using the apparatus to apply measurable forces. After determining κ, the forces can then be determined by monitoring the bead position in the optical potential and applying equation (1.20). Note that the trap stiffness is the second derivative of the parabolic optical potential and is constant for a given setup with fixed parameters and sample, including the chosen trapped bead. It is worth listing here the experimental parameters that strongly influence the value of κ. They are the optical properties of the bead and solvent (refractive indices, absorption coefficient, etc), the bead diameter, the laser power, wavelength and intensity profile, and the characteristics of the objective lens used to focus the laser, especially its numerical aperture [9, 27, 39].

There are a number of different methods currently used to calibrate OT [27, 39]. Here, we will briefly describe one of the simplest of these methods, which is easy to implement in any lab and provides accurate results for the measured κ. Being optically trapped in thermodynamic equilibrium with the surrounding solution, the position probability density of the bead centroid obeys a Boltzmann's distribution. For the x-coordinate, for example,

$$\rho(x) = \frac{1}{\sqrt{2\pi k_B T/\kappa}} \exp\left(\frac{-\kappa x^2}{2k_B T}\right), \tag{1.21}$$

where $\rho(x)$ is the position probability density of the x-coordinate of the bead, k_B is the Boltzmann constant, and T is the absolute temperature.

The variance of the bead x-coordinate in the optical potential is given by

$$\delta x^2 = \langle x^2 \rangle - \langle x \rangle^2 = \langle x^2 \rangle - x_0^2, \tag{1.22}$$

where x_0 if the equilibrium position in the optical potential (see figure 1.2).

The energy equipartition theorem then states that

$$\kappa = \frac{k_B T}{\delta x^2}. \tag{1.23}$$

Therefore, by recording a movie of the bead trapped in the optical potential, one can calculate the variance of the bead position with equation (1.22) and promptly determine the transverse trap stiffness κ using equation (1.23). To perform this, centroid-finding algorithms can be used to analyze the recorded movies and extract the position of the bead center (x, y) as a function of the time.

In order to get accurate results from this type of measurement, it is important to record the movies of the trapped bead with a camera that presents a good resolution and, especially, that allows a high frame rate, capturing on the order of ~1000 fps or higher. This is a fundamental point to determine δx^2 since the Brownian position fluctuations of the bead inside the optical potential are in the scale of milliseconds for a typical trap stiffness on the order of a few pN/μm^{-1} [27, 39].

Finally, the above method can be further improved by correcting the blur effects due to the finite integration time of the cameras, which provides a systematic error in the determination of the centroid of the bead and consequently in the trap stiffness. A useful method to take this task into account is discussed in reference [40].

1.5 Absorbing materials and photophoretic forces

When the bead that one intends to trap is made of a material that considerably absorbs light at the wavelength used in the OT setup, another type of force plays an important role: photophoretic forces, which arise from the asymmetric heat transfer from the bead to the surrounding medium. Figure 1.3 depicts the situation: the laser beam, in yellow, comes from the left and is partially absorbed by the bead, increasing its temperature. Heat then flows from the bead to the surrounding medium, increasing the average kinetic energy of the solvent molecules near the bead surface. The left side of the bead (red colored) tends to transfer more heat because it is in direct contact with the laser beam. Thus, on average, the kinetic energy of the solvent molecules located near this side of the bead will be higher, pushing the bead to the right.

Differently from the two other forces introduced before (scattering and gradient forces), it should be clear that photophoretic forces *are not* optical forces, presenting a thermal origin. Nevertheless, they are fundamental to the bead dynamics in OT when light absorption is non-negligible.

Observe, however, that the net effect of the photophoretic forces is somewhat similar to that of scattering forces, since both tend to push the bead along the

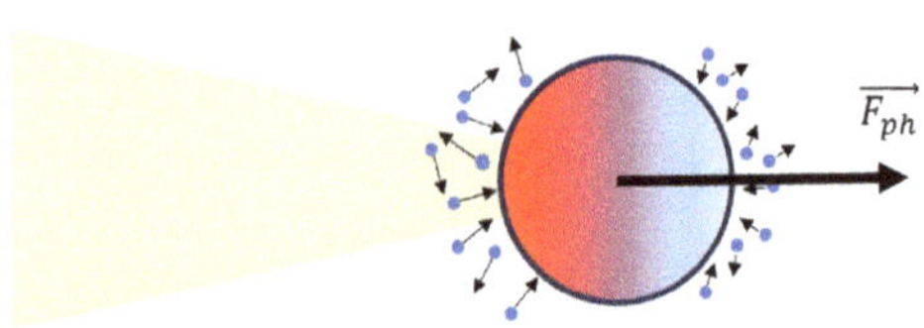

Figure 1.3. Origin of photophoretic forces in absorbing beads: the laser beam, in yellow, comes from the left and is partially absorbed by the bead, increasing its temperature. Heat then flows from the bead to the surrounding medium, increasing the average kinetic energy of the solvent molecules near the bead surface. The left side of the bead (red colored) tends to transfer more heat because it is in direct contact with the laser beam. Thus, on average, the kinetic energy of the solvent molecules located near this side of the bead will be higher, pushing the bead to the right.

direction of the laser propagation, therefore disturbing its stable trapping. Experimentally, it is usually difficult to distinguish between these two types of forces observing only the bead dynamics, as will be discussed in detail in the following chapters. Furthermore, like scattering forces, photophoretic forces are also proportional to the beam intensity. They can usually be written as [41]

$$F_r = C_1 mg(I/r^2) \tag{1.24}$$

Equation (1.24) describes the radiometric force F_r acting on a metallic particle, which arises due to the temperature gradient caused by light absorption. It is proportional to the particle's mass, the gravitational acceleration, and the square of the ratio between the incident light intensity and the particle radius, with a coefficient C_1 that depends on the thermodynamic and hydrodynamic properties of the system.

In any case, the presence of relevant scattering and photophoretic forces is the main reason for the difficulty in trapping conducting micrometer-sized beads, since these materials in general appreciably absorb and reflect light. For metallic particles, for instance, photophoretic forces can be many orders of magnitude higher than for dielectric particles with similar sizes.

The above described situation is eased only for very small Rayleigh particles, because the difference in temperature between the two bead sides will decrease for smaller radii, thus decreasing the photophoretic force. Furthermore, the higher polarizability of metals contributes for higher gradient forces, allowing the stable trapping of very small Rayleigh particles (in practice, with radius of a few tens of nanometers). Therefore, very small Rayleigh metallic particles can usually be stably trapped depending on the setup parameters (laser power, wavelength, waist, etc) or under other special conditions (e.g. special beams such as annular or Bessel beams) [41–44].

And finally, what is the situation for materials with intermediate properties between dielectrics and conductors? What can we predict and observe for the possible different bead dynamics that can arise from such a scenario? What are the possible applications for these types of systems? The main goal of this book is exactly to advance in answering these questions, presenting the state-of-the-art knowledge in the field of optical trapping and manipulation of 'new' (or 'unusual') materials. Please note: the quotation marks are used to refer to these materials as 'new' or 'unusual' because they really are new/unusual for the optical trapping field, although largely studied/used in other fields and technological applications for many decades.

References

[1] Ashkin A 1970 Acceleration and trapping of particles by radiation pressure *Phys. Rev. Lett.* **24** 156

[2] Ashkin A and Dziedzic J M 1987 Optical trapping and manipulation of viruses and bacteria *Science* **235** 1517

[3] Mazolli A, Maia Neto P A and Nussenzveig H M 2003 Theory of trapping forces in optical tweezers *Proc. R. Soc. Lond. Ser. A, Math. Phys. Eng. Sci.* **459** 3021–41
[4] Zhang H and Liu K-K 2008 Optical tweezers for single cells *J. R. Soc. Interface* **5** 671
[5] Fazal F M and Block S M 2011 Optical tweezers study life under tension *Nat. Photon.* **5** 318
[6] Bustamante C, Alexander L, Maciuba K and Kaiser C M 2020 Single-molecule studies of protein folding with optical tweezers *Annu. Rev. Biochem.* **89** 443–70
[7] Bazoni R F, Moura T A and Rocha M S 2020 Hydroxychloroquine exhibits a strong complex interaction with dna: Unraveling the mechanism of action *J. Phys. Chem. Lett.* **11** 9528
[8] Polimeno P, Magazzu A, Iati M A, Patti F, Saija R, Boschi C D E, Donato M G, Gucciardi P G, Jones P H, Volpe G *et al* 2018 Optical tweezers and their applications *J Quant. Spectrosc. Radiat. Transf.* **218** 131–50
[9] Rocha M S 2009 Optical tweezers for undergraduates: theoretical analysis and experiments *Am. J. Phys.* **77** 704
[10] Nieminen T A, Heckenberg N R and Rubinsztein-Dunlop H 2008 Forces in optical tweezers with radially and azimuthally polarized trapping beams *Opt. Lett.* **33** 122
[11] Sakai K and Noda S 2007 Optical trapping of metal particles in doughnut-shaped beam emitted by photonic-crystal laser *Electron. Lett.* **43** 107
[12] Sato S, Harada Y and Waseda Y 1994 Optical trapping of microscopic metal particles *Opt. Lett.* **19** 1994
[13] Shvedov V, Davoyan A R, Hnatovsky C, Engheta N and Krolikowski W 2014 A long-range polarization-controlled optical tractor beam *Nat. Photon.* **8** 846
[14] Ashkin A 1970 Acceleration and trapping of particles by radiation pressure *Phys. Rev. Lett.* **24** 156
[15] Ashkin A 1978 Trapping of atoms by resonance radiation pressure *Phys. Rev. Lett.* **40** 729
[16] Ashkin A, Dziedzic J M, Bjorkholm J E and Chu S 1986 Observation of a single-beam gradient force optical trap for dielectric particles *Opt. Lett.* **11** 288
[17] Ashkin A 1992 Forces of a single-beam gradient laser trap on a dielectric sphere in the ray optics regime *Biophys. J.* **61** 569
[18] Barton J P, Alexander D R and Schaub S A 1988 Internal and near-surface electromagnetic fields for a spherical particle irradiated by a focused laser beam *J. Appl. Phys.* **64** 1632
[19] Barton J P and Alexander D R 1989 Fifth-order corrected electromagnetic field components for a fundamental gaussian beam *J. Appl. Phys.* **66** 2800
[20] Davis L W 1979 Theory of electromagnetic beams *Phys. Rev.* A **19** 1177
[21] Gouesbet G, Maheu B and Gréhan G 1988 Light scattering from a sphere arbitrarily located in a Gaussian beam, using a Bromwich formulation *J. Opt. Soc. Am.* A **5** 1427
[22] Ren K F, Greha G and Gouesbet G 1994 Radiation pressure forces exerted on a particle arbitrarily located in a Gaussian beam by using the generalized Lorenz-Mie theory, and associated resonance effects *Opt. Commun.* **108** 343
[23] Ren K F, Gréhan G and Gouesbet G 1996 Prediction of reverse radiation pressure by generalized Lorenz-Mie theory *Appl. Optics* **35** 2702
[24] Maia Neto P A and Nussenzveig H M 2000 Theory of optical tweezers *Europhys. Lett.* **50** 702
[25] Pesce G, Jones P H, Maragò O M and Volpe G 2020 Optical tweezers: theory and practice *Eur. Phys. J. Plus* **135** 949
[26] Griffiths D J 2017 *Introduction to Electrodynamics* (Cambridge: Cambridge University Press)

[27] Jones P H, Maragò O M and Volpe G 2015 *Optical Tweezers: Principles and Applications* (Cambridge: Cambridge University Press)
[28] Draine B T and Goodman J 1993 Beyond Clausius–Mossotti: Wave propagation on a polarizable point lattice and the discrete dipole approximation *Astrophys. J.* **405** 685
[29] Roosen G 1979 La lévitation optique de sphères *Can. J. Phys.* **57** 1260
[30] Callegari A, Burak Gököz Mite Mijalko A and Volpe G 2015 Computational toolbox for optical tweezers in geometrical optics *J. Opt. Soc. Am.* B **32** B11
[31] Andrade U M S, Garcia A M and Rocha M S 2021 Bessel beam optical tweezers for manipulating superparamagnetic beads *Appl. Opt.* **60** 3422
[32] Gouesbet G and Gréhan G 2011 *Generalized Lorenz-Mie Theories* 1st edn (Berlin: Springer)
[33] Gouesbet G 2010 T-matrix formulation and generalized lorenz-mie theories in spherical coordinates *Opt. Commun.* **283** 517
[34] Richards B and Wolf E 1959 Electromagnetic diffraction in optical systems. ii. structure of the image field in an aplanatic system *Proc. R. Soc. Lond.* A **253** 358–79
[35] Viana N B, Rocha M S, Mesquita O N, Mazolli A and Maia Neto P A 2006 Characterization of objective transmittance for optical tweezers *Appl. Opt.* **45** 4263
[36] Viana N B, Rocha M S, Mesquita O N, Mazolli A, Maia Neto P A and Nussenzveig H M 2007 Towards absolute calibration of optical tweezers *Phys. Rev.* E **75** 021914
[37] Viana N B, Rocha M S, Mesquita O N, Mazolli A, Maia Neto P A and Nussenzveig H M 2006 Absolute calibration of optical tweezers *Appl. Phys. Lett.* **88** 131110
[38] Dutra R S, Viana N B, Maia Neto P A and Nussenzveig H M 2012 Absolute calibration of optical tweezers including aberrations *Appl. Phys. Lett.* **100** 131115
[39] Gieseler J *et al* 2021 Optical tweezers–from calibration to applications: a tutorial *Adv. Opt. Photon.* **13** 74
[40] Wong W P and Halvorsen K 2006 The effect of integration time on fluctuation measurements: calibrating an optical trap in the presence of motion blur *Opt. Exp.* **26** 12517
[41] Ke P C and Gu M 1999 Characterization of trapping force on metallic Mie particles *Appl. Opt.* **38** 160–7
[42] Min C, Shen Z, Shen J, Zhang Y, Fang H, Yuan G, Du L, Zhu S, Lei T and Yuan X 2013 Focused plasmonic trapping of metallic particles *Nat. Commun.* **42891** 2891
[43] Zhan Q 2004 Trapping metallic rayleigh particles with radial polarization *Opt. Exp.* **12** 3377
[44] Huang L, Guo H, Li J, Ling L, Feng B and Li Z 2012 Optical trapping of gold nanoparticles by cylindrical vector beam *Opt. Lett.* **37** 1694
[45] de Oliveira R M, Rocha M S, dos Santos Pires A C and Mendes da Silva L H (ed) 2025 Optical tweezers to probe molecular interactions at the single molecule level *Characterization of Molecular Interactions* (Methods and Protocols in Food Science) (Springer Nature)

IOP Publishing

Optical Trapping and Manipulation of New Materials

Tiago de Assis Moura, Joaquim Bonfim Santos Mendes and Márcio Santos Rocha

Chapter 2

The dipole approximation: Rayleigh regime

In this chapter, we discuss in detail the Rayleigh regime, valid in practice when the particles are much smaller than the laser wavelength used in the optical tweezers (OT). In particular, we discuss how the optical forces can be calculated within this approximation, presenting strategies that allow the development of force models that can be compared to experimental data. The focus is still on dielectric particles, preparing the reader for the calculations that will be performed in the next chapters when dealing with other types of materials.

2.1 Quasi-static approximation

Let us begin our discussion by considering a spherical particle (homogeneous and isotropic) with dielectric constant ε_p, immersed in a dielectric medium (isotropic and non-absorbing) with dielectric constant ε_m, interacting with an electromagnetic (EM) wave, where the wavelength (λ) is much larger than the particle's radius (a) as shown in figure 2.1(a). The phase of the EM wave oscillates harmonically, but due to the condition $\lambda \gg a$, the phase of the EM field remains practically constant over the particle's volume. Thus, we can approximate the spatial distribution of the EM field by assuming the simplified problem of a particle in a uniform electrostatic field ($\mathbf{E} = E_0\hat{z}$). Once the field distributions are determined, the harmonic time dependence can then be incorporated into the solution. This is known as the *quasi-static approximation*, which adequately describes the scattering problem for particles up to 100 nm in size. In this approximation, the electric field is considered uniform throughout the particle volume at any given moment, although it varies over time. Within the particle, the field **E** is spatially constant, while variations may occur on larger scales beyond the particle dimensions. In essence, this approach treats the local field around the particle as a constant electrostatic field.

Consider a dielectric particle, isotropic, non-magnetic ($\mu = 1$) of radius a with dielectric constant ε_p, placed in a uniform electric field $\mathbf{E} = E_0\hat{z}$, as shown in figure 2.1(b). There are no free charges inside the sphere or in the surrounding

doi:10.1088/978-0-7503-6074-6ch2

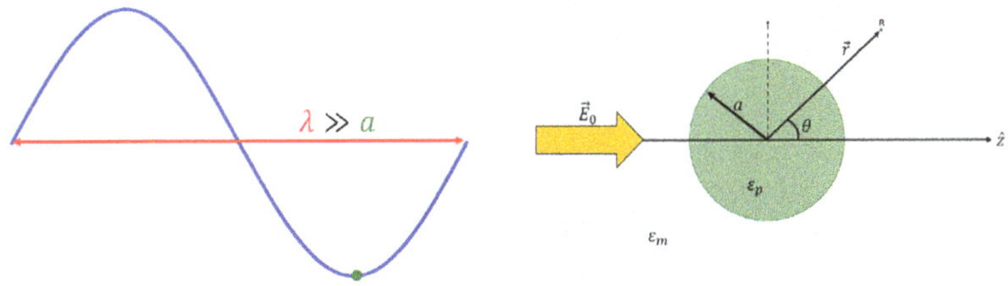

(a) Quasi-static approximation hypothesis.

(b) Sketch of a dielectric particle in a uniform electrostatic field.

Figure 2.1. Dielectric particle with radius a in the dipolar approximation. (a) Diagram illustrating the quasi-static approximation for a plane wave. (b) Sketch showing the geometry of the problem.

medium (with dielectric constant $\varepsilon_{\rm m}$). Therefore, we can solve this problem by solving the Laplace equation ($\nabla^2\Phi = 0$) for the potential Φ ($\mathbf{E} = -\boldsymbol{\nabla}\Phi$) and applying the appropriate boundary conditions at $r = a$. Note that we are taking the position of the particle centroid as the origin of the spherical coordinate frame.

Exploiting the azimuthal symmetry of the problem, the general solution is

$$\Phi(r,\,\theta) = \sum_{l=0}^{\infty}[A_l r^l + B_l r^{-(l+1)}]P_l(\cos\theta). \tag{2.1}$$

Here, $P_l(\cos\theta)$ represents the Legendre polynomials of order l, with θ being the angle between the position vector $\mathbf{r}$ at point R and the z-axis, as illustrated in figure 2.1(b). It is useful to split the solution into two parts: one representing the potential inside the sphere $\Phi_{\rm in}$, and the other representing the potential outside the sphere $\Phi_{\rm out}$. Note that the potential must be finite at $r = 0$, so we can write the solutions as

$$\Phi_{\rm in}(r,\,\theta) = \sum_{l=0}^{\infty} A_l r^l P_l(\cos\theta), \tag{2.2}$$

$$\Phi_{\rm out}(r,\,\theta) = \sum_{l=0}^{\infty}[C_l r^l + D_l r^{-(l+1)}]P_l(\cos\theta). \tag{2.3}$$

The coefficients A_l, C_l and D_l can be obtained from the boundary conditions of the problem:

1. Far from the particle, where $r \gg a$, the potential is given by the potential of the electrostatic constant field $\mathbf{E}_0$.
2. The normal component of $\mathbf{E}$ is discontinuous at $r = a$, but the normal component of the electric displacement is continuous since there are no free charges on the bead surface.
3. The tangential component of $\mathbf{E}$ is continuous at $r = a$.

From the first boundary condition, we have

$$\Phi_{\rm out}(r \gg a,\,\theta) = -E_0 z = -E_0 r\cos\theta. \tag{2.4}$$

Note that this condition is satisfied only if $C_1 = -E_0$ and $C_l = 0$ for $l \neq 1$. So, we obtain all the values of the coefficients C_l. From the second boundary condition, we have

$$-\varepsilon_\mathrm{p}\frac{\partial\Phi_\mathrm{in}}{\partial r}\Big|_{r=a} = -\varepsilon_\mathrm{m}\frac{\partial\Phi_\mathrm{out}}{\partial r}\Big|_{r=a}, \tag{2.5}$$

where we obtain, based on the orthogonality of the Legendre polynomials,

$$A_1 = -\frac{\varepsilon_\mathrm{m}}{\varepsilon_\mathrm{p}}\left(E_0 + 2\frac{D_1}{a^3}\right), \tag{2.6}$$

$$A_l = -\frac{\varepsilon_\mathrm{m}}{\varepsilon_\mathrm{p}}\left[\left(1 + \frac{1}{l}\right)\frac{D_l}{a^{2l+1}}\right], \tag{2.7}$$

where equation (2.7) is valid for $l \neq 1$. From the third boundary condition, we have

$$-\frac{1}{a}\frac{\partial\Phi_\mathrm{in}}{\partial\theta}\Big|_{r=a} = -\frac{1}{a}\frac{\partial\Phi_\mathrm{out}}{\partial\theta}\Big|_{r=a}, \tag{2.8}$$

from where we get

$$A_1 = -E_0 + \frac{D_1}{a^3} \tag{2.9}$$

and

$$A_l = \frac{D_l}{a^{2l+1}}, \tag{2.10}$$

where equation (2.10) is valid for $l \neq 1$. Equations (2.7) and (2.10) are satisfied simultaneously only if $A_l = D_l = 0$ for all $l \neq 1$. By solving the system of equations (2.6) and (2.9), we obtain the coefficients A_1 and D_1 as functions of E_0,

$$A_1 = -\left(\frac{3\varepsilon_\mathrm{m}}{\varepsilon_\mathrm{p} + 2\varepsilon_\mathrm{m}}\right)E_0, \tag{2.11}$$

$$D_1 = \left(\frac{\varepsilon_\mathrm{p} - \varepsilon_\mathrm{m}}{\varepsilon_\mathrm{p} + 2\varepsilon_\mathrm{m}}\right)a^3E_0. \tag{2.12}$$

Thus, we obtain the potential of a sphere interacting with a constant field:

$$\Phi_\mathrm{in} = -\left(\frac{3\varepsilon_\mathrm{m}}{\varepsilon_\mathrm{p} + 2\varepsilon_\mathrm{m}}\right)E_0 r\cos\theta, \tag{2.13}$$

$$\Phi_\mathrm{out} = -E_0 r\cos\theta + \left(\frac{\varepsilon_\mathrm{p} - \varepsilon_\mathrm{m}}{\varepsilon_\mathrm{p} + 2\varepsilon_\mathrm{m}}\right)E_0\frac{a^3}{r^2}\cos\theta. \tag{2.14}$$

From equation (2.13), the electric field inside the particle is given by

$$\mathbf{E}_{\text{in}} = -\nabla\Phi_{\text{in}} = \frac{3\varepsilon_{\text{m}}}{\varepsilon_{\text{p}} + 2\varepsilon_{\text{m}}}\mathbf{E}_0. \tag{2.15}$$

Note that the internal electric field is parallel to the applied electric field. Furthermore, the magnitude of the internal field will be smaller than that of the applied field if the dielectric constant of the particle is higher than that of the surrounding medium. Conversely, if the dielectric constant of the particle is smaller than that of the surrounding medium, the internal field magnitude will exceed that of the applied field. Note that the potential outside the sphere is the sum of the potential due to the applied electric field E_0 and a term arising from the electric field of a dipole, whose the dipole moment **p** (induced in the particle) is given by

$$\mathbf{p} = 4\pi\varepsilon_0\varepsilon_{\text{m}}a^3\left(\frac{\varepsilon_{\text{p}} - \varepsilon_{\text{m}}}{\varepsilon_{\text{p}} + 2\varepsilon_{\text{m}}}\right)\mathbf{E}_0, \tag{2.16}$$

as anticipated in chapter 1.

From this, we write

$$\Phi_{\text{out}} = -E_0 r\cos\theta + \frac{\mathbf{p}.\,\mathbf{r}}{4\pi\varepsilon_0\varepsilon_{\text{m}}r^3}, \tag{2.17}$$

remembering that ε_0 is the electric permittivity of the free space (vacuum). The dipole moment is aligned with the direction of the applied electric field, as expected.

The electric field outside the sphere can be expressed as

$$\mathbf{E}_{\text{out}} = -\nabla\Phi_{\text{out}} = \mathbf{E}_0 + \frac{3\hat{r}(\hat{r}.\,\mathbf{p}) - \mathbf{p}}{4\pi\varepsilon_0\varepsilon_{\text{m}}r^3}, \tag{2.18}$$

where $\hat{r}$ is the unit vector in the direction of the point R of interest. Far from the sphere, the original constant electric field dominates. Figure 2.2(a) illustrates the spatial distribution of the electric field due to the presence of the dielectric sphere.

The volumetric dipole density in the particle can be conveniently represented by the polarization vector **P**, which can be expressed as

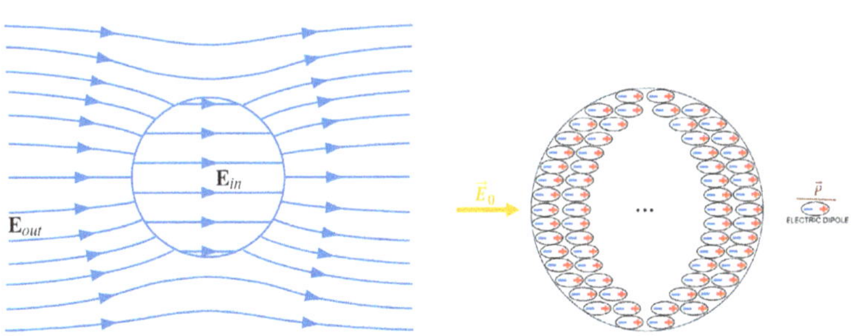

(a) Spatial distribution of an initially uniform electric field around a dielectric particle. (b) Origin of the polarization vector in a dielectric particle.

Figure 2.2. Dielectric particle in an external initially induced electric field. a) Field lines in the presence of the dielectric sphere. b) Origin of the polarization vector.

$$\mathbf{P} = 3\varepsilon_0\varepsilon_m\left(\frac{\varepsilon_p - \varepsilon_m}{\varepsilon_p + 2\varepsilon_m}\right)\mathbf{E}_0. \tag{2.19}$$

But what is the origin of **P**? Although the dielectric particle has no net electric charge, the application of an external electric field $\mathbf{E}_0$ causes a separation of charges within the particle. This separation occurs because the center of mass of the positive charges (atomic nuclei) and the center of mass of the negative charges (electron cloud) within the atoms or molecules do not perfectly coincide in response to the applied field. This displacement generates electric dipoles at the atomic or molecular level. The polarization vector **P** represents the volumetric density of these aligned dipoles, as can be seen in figure 2.2(b).

With the application of $\mathbf{E}_0$, the induced dipoles align in the direction of the field. Inside the particle, the dipoles are arranged in such a way that their internal charges cancel each other out, resulting in a net zero charge within the volume of the particle. However, at the surface of the particle, there is a misalignment between the ends of the dipoles, leading to an accumulation of surface charges. As a result, one side of the particle will have an excess of negative charges, while the other side will have an excess of positive charges, thereby creating an external dipole field.

These polarization surface charges arise from the discontinuity in the normal component of the polarization vector at the interface between the medium and the particle and can be written as [1]

$$\sigma_{pol} = -(\mathbf{P}_{out} - \mathbf{P}_{in}) \cdot \hat{n} = 3\varepsilon_0\left(\frac{\varepsilon_p - \varepsilon_m}{\varepsilon_p + 2\varepsilon_m}\right)E_0 \cos\theta, \tag{2.20}$$

where $\hat{n}$ is the normal unit vector, $\mathbf{P}_{out} = \varepsilon_0(\varepsilon_m - 1)\mathbf{E}_{out}$ and $\mathbf{P}_{in} = \varepsilon_0(\varepsilon_p - 1)\mathbf{E}_{in}$. Note that the same result can be obtained by considering the discontinuity of the electric field, given that the free charge density is zero. In this case, we have $\sigma = \sigma_{pol}$, leading to

$$\sigma_{pol} = \varepsilon_0(\mathbf{E}_{out} - \mathbf{E}_{in}) \cdot \hat{n} = 3\varepsilon_0\left(\frac{\varepsilon_p - \varepsilon_m}{\varepsilon_p + 2\varepsilon_m}\right)E_0 \cos\theta. \tag{2.21}$$

Induced charges play a fundamental role in optical traps, so let us delve deeper in analyzing them. Since we are assuming that our dielectric sphere is immersed in a non-absorptive medium with a dielectric constant ε_m, the distribution of charges on the surface of the sphere depends on the relationship between the dielectric constants of the sphere and the surrounding medium. Figure 2.3 illustrates the relationship between the dielectric constants of the medium and the particles, along with the resulting charge distributions due to the interaction with a uniform electric field $\mathbf{E}_0$. In panel (a) the dielectric constant of the particle (ε_1) is higher than that of the medium, while in panel (b) the dielectric constant of the particle (ε_2) is smaller than that of the medium.

When $\varepsilon_1 > \varepsilon_m$ the polarization vector **P** aligns with the external field $\mathbf{E}_0$, leading to an accumulation of positive charges at the north pole ($\theta = 0$) and negative

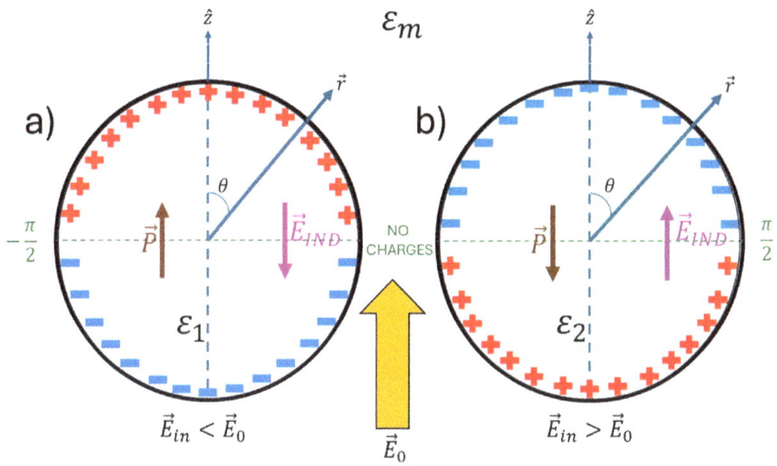

Figure 2.3. Surface charge density induced on a dielectric particle due to a uniform electric field. (a) Case when the dielectric constant of the sphere is higher than that of the surrounding medium. (b) Case when the dielectric constant of the sphere is smaller than that of the surrounding medium.

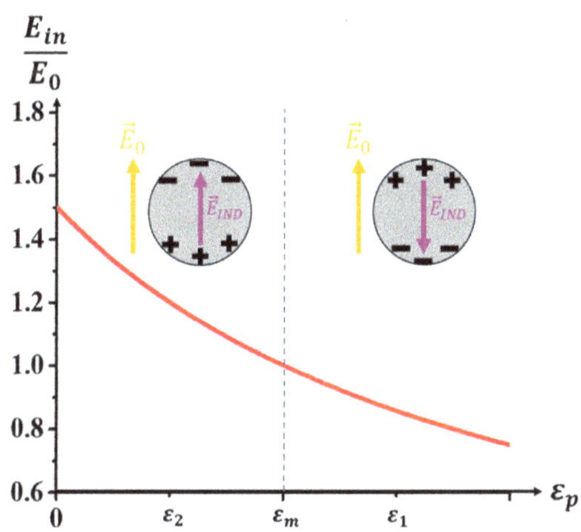

Figure 2.4. Diagram illustrating the variation of the electric field inside the particle as a function of its dielectric constant. ε_1 and ε_2 are equivalent to those in figure 2.3.

charges at the south pole ($\theta = \pi$) of the sphere. This charge separation generates an induced field ($\mathbf{E}_{\text{ind}}$) within the sphere that opposes the incident field. As a result, the net field inside the sphere ($\mathbf{E}_{\text{in}}$) is reduced compared to $\mathbf{E}_0$.

On the other hand, when $\varepsilon_2 < \varepsilon_m$, the situation is reversed. Negative charges now gather at the north pole, with positive charges at the south pole, causing the polarization vector to oppose the external field. In this configuration, the induced field ($\mathbf{E}_{\text{ind}}$) aligns with $\mathbf{E}_0$, amplifying the internal field ($\mathbf{E}_{\text{in}}$), which becomes stronger than the incident field. Note that the internal field $\mathbf{E}_{\text{in}}$ is always in phase with the incident field $\mathbf{E}_0$, regardless of the ratio $\varepsilon_p/\varepsilon_m$. Figure 2.4 illustrates the behavior of the ratio E_{in}/E_0 as a function of the dielectric constant of the dielectric sphere ε_p.

Thus, the nature of the material, whether its dielectric constant is higher or smaller than the surrounding medium, directly dictates on how the induced charges respond to the applied field, decreasing or enhancing the internal electric field within the sphere. In both scenarios, there is no charge density along the equator of the sphere ($\theta = \pm\pi/2$). This occurs because, at the equatorial region, both the electric field and the polarization vector remain continuous across the interface between the medium and the sphere. As a result, no net surface charges accumulate in this region, ensuring charge neutrality along the equator.

An important result that can be deduced from equation (2.16) is the polarizability (α) of the sphere, which is defined in terms of the dipole moment as

$$\mathbf{p} = \varepsilon_0 \varepsilon_{\mathrm{m}} \alpha \mathbf{E}_0, \tag{2.22}$$

where

$$\alpha = 4\pi a^3 \frac{\varepsilon_{\mathrm{p}} - \varepsilon_{\mathrm{m}}}{\varepsilon_{\mathrm{p}} + 2\varepsilon_{\mathrm{m}}}, \tag{2.23}$$

as anticipated in chapter 1.

Physically, polarizability is a quantity that relates the dipole moment induced in the sphere to the applied electric field that generated this dipole moment. In other words, the polarizability (α) quantifies how easily a material develops an electric dipole moment (**p**) when exposed to an external electric field. The polarization vector (**P**) describes the density of dipole moments per unit volume within the material and should not be confused with the dipole moment itself. In summary, polarizability governs the material response to the electric field, resulting in an induced dipole moment, and the polarization vector characterizes this response at a macroscopic scale. Equation (2.23) is known, in the static limit, as the Clausius–Mossotti relation. When $\varepsilon_{\mathrm{p}} > \varepsilon_{\mathrm{m}}$, we have $\alpha > 0$, and when $\varepsilon_{\mathrm{p}} < \varepsilon_{\mathrm{m}}$, we find $\alpha < 0$, reflecting the dependence of polarizability on the distribution of surface charges.

So far we have considered the interaction of a dielectric sphere with an uniform static field. However, our objective is to discuss its interaction with an EM field that varies harmonically with time. The treatment of the dielectric particle as a dipole remains valid for an EM field as long as $\lambda \gg a$. In other words, within the quasi-static approximation, the problem of a spherical particle in an EM field can be effectively reduced to that of an electric dipole interacting with the field.

Under plane-wave illumination, where $\mathbf{E}_i(\mathbf{x}, t) = \mathbf{E}_0 e^{i(\mathbf{k}\cdot\mathbf{x}-\omega t)}$ and $\mathbf{H}_i(\mathbf{x}, t) = \mathbf{E}_i/\mu_0 c$, the induced dipole moment on the sphere also exhibits harmonic time dependence and can be expressed as $\mathbf{p}(t) = \varepsilon_0 \varepsilon_{\mathrm{m}} \alpha \mathbf{E}_0(\mathbf{r}) e^{-i\omega t}$, where α is the static polarizability given by equation (2.23). As usual, the real part of such expressions is to be taken to obtain physical quantities.

Note that the polarization vector, and consequently the surface charge density, exhibit time dependence. This temporal variation in the induced surface charges can be understood as a natural response of the charges to the oscillation of the applied electric field. The modulation of the electric field amplitude over time induces corresponding oscillations in the material polarization.

Unlike in metals, where the induced charges have a response time compatible with the oscillation of the electric field at optical frequencies, in dielectrics the response time of the induced charges is much shorter compared to the field oscillation. As a result, for the charges to oscillate at the field frequency, such oscillation is forced. This is in contrast with metals, where the oscillations are resonant. We evaluate the effects of the resonance between the induced charges and the incident field in chapter 4.

As shown in figure 2.3, the distribution of induced charges depends on the relationship between the dielectric constant of the sphere and that of the surrounding medium. Figure 2.5 illustrates the variation in the distribution of surface-induced charges as a function of the time for a plane wave, with $\varepsilon_1 > \varepsilon_m$ in figure 2.5(a), and $\varepsilon_2 < \varepsilon_m$ in figure 2.5(b).

As in the static case, if $\varepsilon_p > \varepsilon_m$, the polarization **P** is always parallel to the incident field, whereas if $\varepsilon_p < \varepsilon_m$, **P** is always antiparallel to the incident field. Therefore, when the amplitude of the incident electric field changes sign, **P** follows this change to remain parallel or antiparallel to the incident field, leading to oscillations in the induced surface charge distribution. In other words, under an oscillating electric field, the dielectric sphere behaves like an oscillating dipole.

Equation (2.23) can be improved to account for the fact that the dielectric constant of the particle will be a function of the wavelength in the presence of an EM wave. In general, such a constant will be a complex function, reflecting the cases where the sphere absorbs part of the incident radiation:

$$\alpha(\lambda) = 4\pi a^3 \frac{\varepsilon_p(\lambda) - \varepsilon_m}{\varepsilon_p(\lambda) + 2\varepsilon_m}. \tag{2.24}$$

Equation (2.24) is known as the Lorentz–Lorenz relation, and the dependence of ε_p on λ will be deduced below. For simplicity let us consider a non-absorbing medium such that ε_m is a positive real number. In contrast, $\varepsilon_p = \varepsilon' + i\varepsilon'$ is complex, where ε' is related to evanescent wave propagation in the material.

The complex nature of ε_p leads to a complex polarizability. The real part of the polarizability represents the oscillation of the dipole in phase with the EM field

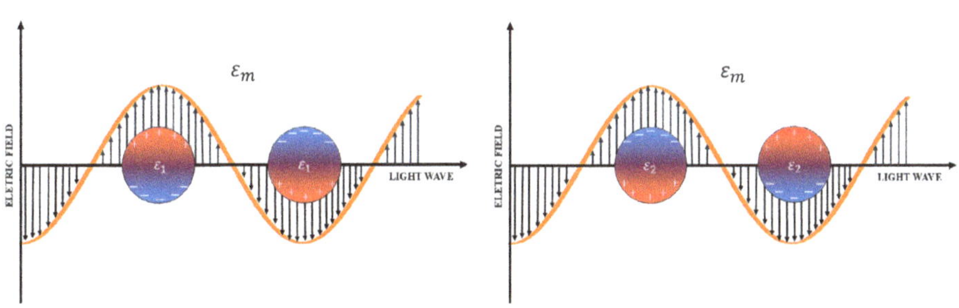

(a) Oscillation of induced charges for $\varepsilon_1 > \varepsilon_m$. (b) Oscillation of induced charges for $\varepsilon_2 < \varepsilon_m$.

Figure 2.5. Oscillation of the induced charge density in response to the oscillation of the electric field. (a) Case where $\varepsilon_p = \varepsilon_1 > \varepsilon_m$. (b) Case where $\varepsilon_p = \varepsilon_2 < \varepsilon_m$. ε_1 and ε_2 are equivalent to those in figure 2.3.

(if $\varepsilon' > \varepsilon_m$) and out of phase (with a phase shift of π, if $\varepsilon' < \varepsilon_m$). In contrast, the imaginary part of the polarizability reflects oscillation in quadrature phase (with a phase shift of $\pi/2$).

Before we continue, let us briefly review the theory associated with an oscillating electric dipole. For readers interested in a more in-depth discussion, we recommend consulting chapters 9 and 10 of reference [1].

Note that the oscillatory nature of the surface charge distribution leads to the emergence of polarization currents (**J**). These polarization currents act as sources of electric fields **E** and magnetic fields **H**. Let us consider $\mathbf{x}'$ as the coordinates of the source, which, for simplicity, we will assume to be in the free space. The fields **E** and **H** can be obtained through the vector potential **A**. Using the Lorenz gauge we have

$$\mathbf{A}(\mathbf{x}) = \frac{\mu_0}{4\pi}\int d^3x' \int dt' \frac{\mathbf{J}(\mathbf{x}', t')}{|\mathbf{x} - \mathbf{x}'|}\delta\left(t' + \frac{|\mathbf{x} - \mathbf{x}'|}{c} - t\right), \tag{2.25}$$

where c is the speed of light and the Dirac delta function ensures the causal behavior of the fields. With harmonic time dependence, equation (2.25) becomes

$$\mathbf{A}(\mathbf{x}) = \frac{\mu_0}{4\pi}\int \mathbf{J}(\mathbf{x}')\frac{e^{ik|\mathbf{x}-\mathbf{x}'|}}{|\mathbf{x} - \mathbf{x}'|}d^3x', \tag{2.26}$$

where $k = \omega/c$ is the wave number. Again, the time dependence is implied. Once we know $\mathbf{A}(\mathbf{x})$, the field **H** can be expressed as

$$\mathbf{H} = \frac{1}{\mu_0}\nabla \times \mathbf{A}, \tag{2.27}$$

while outside the sphere the electric field is

$$\mathbf{E} = \frac{iZ_0}{k}\nabla \times \mathbf{H}, \tag{2.28}$$

where $Z_0 = \sqrt{\mu_0/\varepsilon_0}$ is the impedance of the free space.

Thus, knowing the current distribution $\mathbf{J}(\mathbf{x}')$, the fields **H** and **E** can be obtained using equations (2.26), (2.27), and (2.28). The task then reduces to calculating the fields. However, it is possible to derive a simpler, yet general, relation for the field properties in the limit where the current sources are confined to a small sphere of radius a, assuming $a \ll \lambda$, which is the case of interest here.

Note that

$$k = \frac{2\pi}{\lambda}.$$

Let r be the distance between the source and the point where we want to obtain the fields, then

$$kr = \frac{2\pi r}{\lambda} \sim \frac{r}{\lambda}.$$

Since we are interested in the case where $a \ll \lambda$, there are three spatial regions of interest, with the fields exhibiting very different behaviors in each region.

1. The near (static) zone: $a \ll r \ll \lambda$ or $kr \ll 1$.
2. The intermediate (induction) zone: $a \ll r \sim \lambda$.
3. The far (radiation) zone: $a \ll \lambda \ll r$ or $kr \gg 1$.

In the near zone the fields exhibit the characteristics of static fields, with radial components and variations in distance that depend on the detailed properties of the source. In the far zone, however, the fields are transverse to the radius vector and decay with r^{-1}, which is typical of radiation fields; hence, this region is also called the radiation zone. In the intermediate zone the calculations become more complex, requiring a multipole expansion of the fields. Figure 2.6 provides a schematic diagram illustrating these zones.

Here, we are primarily interested in the behavior of the fields in the radiation zone ($kr \gg 1$); the analysis of near-field will be done in chapter 4. In the radiation zone, the exponential term in equation (2.26) oscillates rapidly and governs the behavior of $\mathbf{A}$. Since $a \ll r$, we can use the following approximation:

$$|\mathbf{x} - \mathbf{x}'| \simeq r - \hat{e}_x \cdot \mathbf{x}', \tag{2.29}$$

where $\hat{e}_x$ is a unit vector in the direction of $\mathbf{x}$. Substituting equation (2.29) into equation (2.26) and retaining only the dominant term, the vector potential can be expressed as

$$\lim_{kr\to\infty} \mathbf{A}(\mathbf{x}) = \frac{\mu_0}{4\pi}\frac{e^{ikr}}{r}\int \mathbf{J}(\mathbf{x}')e^{-ik\hat{e}_x \cdot \mathbf{x}'}d^3x'. \tag{2.30}$$

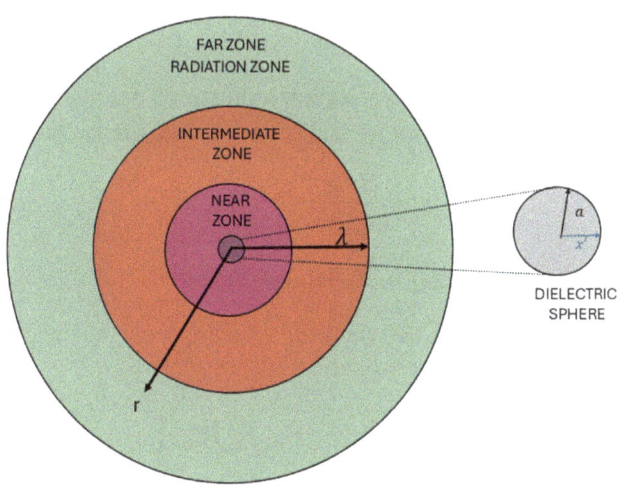

Figure 2.6. Diagram illustrating the spatial regions (zones) where EM fields exhibit different characteristics. In purple the near zone, where the fields display static-like properties. In orange, the intermediate zone. In green, the far zone or radiation zone, where the fields are described as radiation fields.

Equation (2.30) demonstrates that, in the far zone, the vector potential behaves like a spherical wave propagating outward from the sphere. Using equations (2.27) and (2.28), it is straightforward to show that the **E** and **H** are transverse to the radial vector and decay as r^{-1}, thus corresponding to radiation fields. Since the dimensions of the source are smaller than the wavelength, we can expand the three-dimensional integral in powers of k:

$$\lim_{kr\to\infty} \mathbf{A}(\mathbf{x}) = \frac{\mu_0}{4\pi}\frac{e^{ikr}}{r}\sum_n \frac{(-ik)^n}{n!}\int \mathbf{J}(\mathbf{x}')(\hat{e}_x \cdot \mathbf{x}')^n d^3x'. \tag{2.31}$$

The higher-order terms in equation (2.31) are generally negligible, meaning the radiation emitted by the source arises primarily from the first non-zero term of the expansion. By retaining only the first term in equation (2.31), the vector potential becomes

$$\mathbf{A}(\mathbf{x}) = \frac{\mu_0}{4\pi}\frac{e^{ikr}}{r}\int \mathbf{J}(\mathbf{x}')d^3x'. \tag{2.32}$$

Using integration by parts:

$$\int \mathbf{J}d^3x' = -\int \mathbf{x}'(\boldsymbol{\nabla} \cdot \mathbf{J})d^3x', \tag{2.33}$$

from the continuity equation we have

$$\frac{\partial \rho}{\partial t} = i\omega\rho = \boldsymbol{\nabla} \cdot \mathbf{J}. \tag{2.34}$$

Thus the vector potential is

$$\mathbf{A}(\mathbf{x}) = -\frac{i\mu_0\omega}{4\pi}\frac{e^{ikr}}{r}\mathbf{p}, \tag{2.35}$$

where $\mathbf{p} = \int \mathbf{x}'\rho(\mathbf{x}')d^3x'$ is the electric dipole moment as defined in the electrostatics case by (2.16). Then, using equations (2.27) and (2.28), the radiation fields are

$$\mathbf{H}_{\text{rad}} = \frac{ck^2}{4\pi}\frac{e^{ikr}}{r}(\hat{r} \times \mathbf{p}), \tag{2.36}$$

$$\mathbf{E}_{\text{rad}} = Z_0(\mathbf{H}_{\text{rad}} \times \hat{r}). \tag{2.37}$$

The time-average power radiated per unit solid angle by the dielectric sphere with dipole moment **p** is

$$\frac{dP_{\text{rad}}}{d\Omega} = \frac{1}{2}\text{Re}\left[r^2\hat{r} \cdot \mathbf{E}_{\text{rad}} \times \mathbf{H}^*_{\text{rad}}\right]. \tag{2.38}$$

Using equations (2.36) and (2.37) in equation (2.38), we obtain

$$\frac{dP_{\text{rad}}}{d\Omega} = \frac{c^2 Z_0}{32\pi^2}k^4|(\hat{r} \times \mathbf{p}) \times \hat{r}\,|^2. \tag{2.39}$$

When the components of $\mathbf{p}$ all have the same phase, the angular distribution exhibits the typical configuration of a dipole,

$$\frac{dP_{\text{rad}}}{d\Omega} = \frac{c^2 Z_0}{32\pi^2} k^4 |\mathbf{p}|^2 \sin^2\theta. \tag{2.40}$$

Figure 2.7 shows the distribution of the angular power radiated by a dipole, as given by equation (2.40). Note that there is no radiated power along the dipole axis ($\theta = 0$ and $\theta = \pi$), while the power is maximized in the plane perpendicular to the dipole axis, resulting in an anisotropic angular distribution. Additionally, there is a dependence on k^4 (known as Rayleigh's Law), which is an almost universal characteristic of the radiation scattering by objects whose dimensions are much smaller than the wavelength of the incident radiation. This implies that the radiated power increases significantly with the increasing frequency of the incident wave. This dependence becomes more evident when calculating the total radiated power (which is independent of the relative phases of the $\mathbf{p}$ components). By integrating equation (2.40) we obtain

$$P_{\text{rad}} = \int \frac{dP_{\text{rad}}}{d\Omega} d\Omega = \frac{c^2 Z_0}{12\pi} k^4 |\mathbf{p}|^2. \tag{2.41}$$

An important consequence of the generation of $\mathbf{E}$ is that it also influences the polarization of the sphere, in addition to the external field. The inclusion of $\mathbf{E}$ (mainly the term that corresponds to the near field) in the calculation of $\mathbf{P}$ introduces

Figure 2.7. Distribution of the angular power radiated by a dipole, as given by equation (2.40). Note that there is no radiated power along the dipole axis ($\theta = 0$ and $\theta = \pi$), while the power is maximized in the plane perpendicular to the dipole axis, resulting in an anisotropic angular distribution.

a self-interaction term, which leads to corrections in the Clausius–Mossotti relation (α_{CM}, given by equation (2.23) taking $\varepsilon_m = 1$).

These corrections are particularly useful in non-absorbing dielectrics for computing losses due to scattering and/or absorption of the incident wave. Furthermore, this correction is essential for understanding the effects of localized surface plasmon (LSP) resonance on optical forces for conducting particles, as we will discuss in chapter 4. The introduction of the radiation field into the polarization vector leads to the construction of an effective polarizability α_{rad}, given by[1]:

$$\alpha_{rad} = \frac{\alpha_{CM}}{1 - \frac{\varepsilon_p - 1}{\varepsilon_p + 2}\left[(ka)^2 + \frac{2i}{3}(ka)^3\right]}, \tag{2.42}$$

a formula anticipated in chapter 1.

In this case, equation (2.42) is evaluated for a sphere in a vacuum. Notably, even for a dielectric where ε_p is a real number, α_{rad} will have an imaginary component. This imaginary part corresponds to the dissipation of energy in the form of radiation or absorption by the dielectric. This loss of energy from the incident wave due to scattering or absorption indicates that the induced dipole is in quadrature phase with the electric field of the incident wave. Thus, the dielectric sphere not only polarizes but also radiates (or absorbs) energy, leading to a transfer of energy from the incident field to the surrounding medium.

Equation (2.42) can be simplified by neglecting the term $(ka)^2$ in the denominator and generalized to a surrounding (non-absorbing) medium with dielectric constant ε_m,

$$\alpha_{rad} = \frac{\alpha_{CM}}{\left[1 - \frac{ik_m^3\alpha_{CM}}{6\pi}\right]}, \tag{2.43}$$

where $k_m = \sqrt{\varepsilon_m}k$ is the wave number in the surrounding medium.

Let us now reconsider what we just discussed from a different perspective, which, as we will see, will be much more useful throughout this chapter. Initially, all the energy is stored in the fields of the incident wave ($\mathbf{E}_i$, $\mathbf{H}_i$), which can be quantified by the intensity of the incident wave in a non-absorbent medium with dielectric constant ε_m, given by[2]

---

[1] Details of the derivation of the effective polarizability can be found in reference [2].

[2] The reader may have noticed that the intensity of an EM wave in a non-magnetic and non-absorbing material medium is slightly higher than in the vacuum, according to $I_{material} = \sqrt{\varepsilon_m} I_{vacuum}$. This increase in intensity occurs because the material can store more electrical energy in the wave field, which is directly related to its electrical permittivity, ε_m.
Physically, this is due to the interaction of the wave electric field with the material medium, which reduces the wave propagation speed. This reduction in the propagation speed (by a factor of $1/\sqrt{\varepsilon_m}$) means that the wave energy is distributed over a smaller spatial region along its path, increasing its energy density. Additionally, the factor ε_m increases the electrical energy stored in the field. The combination of these two effects results in a higher intensity, which is reflected by the Poynting vector in the medium, amplified by a factor of $\sqrt{\varepsilon_m}$ compared to vacuum ($S_{material} = \sqrt{\varepsilon_m} S_{vacuum}$).

$$I = |\langle \mathbf{S}_{\mathrm{i}} \rangle| = \frac{1}{2}|\mathrm{Re}\langle \mathbf{E}_i \times H_i^* \rangle| = \frac{1}{2}\sqrt{\frac{\varepsilon_0 \varepsilon_\mathrm{m}}{\mu_0}}E_0^2 = \frac{1}{2}c\sqrt{\varepsilon_\mathrm{m}}\varepsilon_0 E_0^2, \tag{2.44}$$

where $\mathbf{S}_i = \mathrm{Re}(\mathbf{E}_i) \times \mathrm{Re}(\mathbf{H}_i)$ is the Poynting vector of the incident wave, $\langle \mathbf{S}_\mathrm{i} \rangle$ is the temporal average of $\mathbf{S}_\mathrm{i}$ and E_0 is the amplitude of the electric field of the incident wave. The intensity can be interpreted as the amount of energy that passes through a unit area perpendicular to the direction of propagation of the wave per unit of time.

When an EM wave interacts with a dielectric sphere much smaller than its wavelength ($a \ll \lambda$), a portion of the wave energy is converted into electric potential energy, creating an induced electric dipole within the sphere. This dipole oscillates in sync with the incident electric field—either in phase or out of phase by π, depending on the relationship between ε_p and ε_m. The work required to sustain these charge oscillations is done at the expense of the energy of the EM wave, which transfers a fraction of its energy to the sphere, where it is temporarily stored in the oscillating dipole. Since the polarized dielectric sphere acts as an oscillating electric dipole, it re-emits the absorbed energy back into the surrounding medium. In the far zone ($k_\mathrm{m} r \gg 1$), this energy is emitted as EM radiation through the radiation fields $\mathbf{E}_\mathrm{rad}$ and $\mathbf{H}_\mathrm{rad}$ in all directions in the form of a spherical wave.

In other words, during the interaction of the incident wave with the dielectric sphere, the energy contained in the wave fields is redistributed among several processes. A portion of this energy is stored in the fields associated with the sphere polarization, referred to as electrostatic energy. Another portion is transferred to the oscillatory motion of surface charges, which generates induced currents. These currents, in turn, re-emit EM radiation, whose intensity and angular distribution depend on the optical properties of the material and the interaction conditions. In non-ideal materials, an additional fraction of the energy may be dissipated as heat due to losses associated with the material conductivity or dielectric characteristics.

We define *extinction* as the fraction of energy removed from the incident wave, regardless of its subsequent conversion. Extinction, therefore, refers to the effective reduction in the intensity of the EM wave in the direction of its propagation. This removed energy includes both the scattered energy, re-emitted in the form of radiation with a predominantly radial Poynting vector, and the absorbed energy that can be converted into heat or other forms of energy.

To quantify the extinction we use Poynting's theorem, which allows us to determine the average rate at which energy is removed from the incident wave fields. The extinction power, P_ext, is calculated by the average work done by the incident electric field, $\mathbf{E}_i$, on the induced current density, $\mathbf{J}$.

At optical frequencies, on the order of 10^{15} Hz, the instantaneous values of the fields vary extremely rapidly, making it impractical to track these variations in real time. Moreover, measurement instruments record time-averaged values, and the physical effects of interest, such as optical forces and energy transfer, are better described in terms of time averages. For this reason, only the temporal average, calculated over a full oscillation period, is relevant for describing observable physical effects. Thus, P_ext is given by

$$P_{\text{ext}} = \frac{1}{2}\int_V \text{Re}\{\mathbf{J} \cdot \mathbf{E}_i^*\} dV. \tag{2.45}$$

The induced current density around the oscillating dipole can be calculated as

$$\mathbf{J} = -i\omega\mathbf{p}\delta(\mathbf{r} - \mathbf{r}_d). \tag{2.46}$$

Thus, we can write equation (2.45) as

$$P_{\text{ext}} = \frac{\omega}{2} Im\{\mathbf{p} \cdot \mathbf{E}_i^*(\mathbf{r}_d)\}. \tag{2.47}$$

Here, $\mathbf{p}$ is the dipole moment and $\mathbf{E}_i$ is evaluated at the origin of the dipole $\mathbf{r}_d$, (i.e. at the centroid of the sphere). Writing the dipole moment as a function of the radiative polarizability (equation (2.43)), we have

$$P_{\text{ext}} = \frac{\omega}{2}\varepsilon_0\varepsilon_{\text{m}}|\mathbf{E}_i|^2 Im\{\alpha_{\text{rad}}\}. \tag{2.48}$$

We define the extinction cross-section (σ_{ext}) as the ratio of the extinguished power to the intensity of the incident wave. In other words, σ_{ext} represents an effective area through which the particle interacts with the incident wave, reducing its intensity in the direction of propagation,

$$\sigma_{\text{ext}} = \frac{P_{\text{ext}}}{I} = k_{\text{m}} Im\{\alpha_{\text{rad}}\}. \tag{2.49}$$

Similarly, we can define the scattering cross-section (σ_{sct}), which represents the effective area over which the dielectric sphere acts as a scattering center for the incident wave. This quantity measures the fraction of the incident wave energy that is redirected in different directions due to scattering by the particle[3], without significant absorption of energy. In other words, the scattering cross-section describes the portion of the incident energy that is absorbed and immediately re-emitted in the form of radiation, contributing to the redistribution of the wave energy around the particle. Thus, mathematically, we can define the scattering cross-section, σ_{sct}, as the ratio between the power scattered by the sphere, P_{sct}, and the intensity of the incident wave, I,

$$\sigma_{\text{sct}} = \frac{P_{\text{sct}}}{I}. \tag{2.50}$$

The reader should recall that we have already obtained the value of the power scattered by the dielectric sphere by calculating the radiation fields $\mathbf{H}_{\text{rad}}$ and $\mathbf{E}_{\text{rad}}$ in equations (2.36), (2.37), and (2.41). However, until now, we have not explained how equation (2.38) allows us to obtain the radiated power—which, in this context, is synonymous with scattering. Let us now clarify how equation (2.38) leads to the calculation of the power scattered by the dielectric sphere.

[3] Here we are considering only elastic scattering, in which the radiated energy has the same frequency of the incident energy.

Note that equation (2.38) is a direct consequence of the principle of conservation of energy. To verify this, consider a portion of space defined by a volume V bounded by its surface ∂V. If the EM energy, u_{em}, varies within V, then either the energy is flowing across the surface ∂V, or it is being converted into other forms of energy within the volume (which we will define as u_{mat}). Using the continuity equation for EM energy, we have

$$\int_V \frac{\partial u_{\text{em}}}{\partial t} dV - \oint_{\partial V} \langle \mathbf{S} \rangle \cdot \hat{n} da = -\int_V \frac{\partial u_{\text{mat}}}{\partial t} dV, \tag{2.51}$$

where the left side represents the variation of EM energy, and the right side represents the variation of energy in matter. It is worth noting that both u_{em} and $\mathbf{S}$ in equation (2.51) represent the energy density and the total energy flux present in all EM fields of the system, including the fields of the incident wave as well as those of radiation, polarization, and other contributions. Defining the total energy density contained in the volume V as $u_{\text{t}} = u_{\text{em}} + u_{\text{mat}}$ and rearranging the terms of equation (2.51), we have

$$\int_V \frac{\partial u_{\text{t}}}{\partial t} dV = \oint_{\partial V} \langle \mathbf{S} \rangle \cdot \hat{n} da. \tag{2.52}$$

Equation (2.52) indicates that the variation of the total energy contained in the volume V can only occur if there is a flow of energy crossing the surface that delimits this volume (the boundary ∂V). Since the energy contained in the matter is confined to it (and, by hypothesis, the volume V encompasses all the matter), the flow of energy through ∂V can only be associated with the EM energy, represented by the Poynting vector $\mathbf{S}$.

Since the surface ∂V is arbitrary, and we assume that the medium is non-absorbing, we can take the limit where ∂V extends to infinity. In this scenario, only the fluxes associated with the incident fields and the radiated fields are relevant for calculating the flux of $\mathbf{S}$. Thus, we can write $\mathbf{S} = \mathbf{S}_{\text{i}} + \mathbf{S}_{\text{scat}}$, where $\mathbf{S}_{\text{i}}$ represents the flux of the incident wave, and $\mathbf{S}_{\text{scat}}$ represents the flux associated with the scattered radiation. Substituting this decomposition into equation (2.52) we obtain

$$\int_V \frac{\partial u_{\text{t}}}{\partial t} dV = \oint_{\partial V} \langle \mathbf{S}_{\text{i}} \rangle \cdot \hat{n} da + \oint_{\partial V} \langle \mathbf{S}_{\text{scat}} \rangle \cdot \hat{n} da. \tag{2.53}$$

For a plane wave, the integral of the flow of $\mathbf{S}_i$ is zero[4], causing all the information on the variation of u_{t} to be computed by the integral of the flow of $\langle \mathbf{S}_{\text{scat}} \rangle$, which is finite and non-zero

$$\int_V \frac{\partial u_{\text{t}}}{\partial t} dV = \oint_{\partial V \to \infty} \langle \mathbf{S}_{\text{scat}} \rangle \cdot d\mathbf{a}. \tag{2.54}$$

[4] The surface integral of the flow of the Poynting vector of the incident wave $\mathbf{S}_{\text{i}}$ over a closed area tending to infinity and encompassing the dielectric sphere vanishes, because for a plane wave, the flux of $\mathbf{S}_{\text{i}}$ is uniform and constant along the surface. The energy of the incident wave enters on one side of the surface and leaves symmetrically on the opposite side, without producing any net flux. That is, in the absence of the dielectric sphere, the total energy (which is just the energy stored in the fields of the incident wave quantified by u_{em}) is conserved, as expected.

Since the flow of $\langle \mathbf{S}_{\text{scat}} \rangle$ does not vanish in the limit where the surface ∂V tends to infinity, the total energy u_t is not conserved in this volume. This indicates the existence of an energy flux leaving the surface, originating from the EM waves emitted by the dielectric sphere in the form of oscillating dipole radiation.

Dipole radiation fields decay as $1/r$, which implies that $\langle \mathbf{S}_{\text{rad}} \rangle$ decays as $1/r^2$. Thus, the energy flow $\oint_{\partial V} \langle \mathbf{S}_{\text{rad}} \rangle \cdot \hat{n}\, da$, remains constant as $r \to \infty$, reflecting the conservation of energy in the system.

Applying Poynting's theorem, we can associate the flow of $\langle \mathbf{S}_{\text{scat}} \rangle$ with the average rate of work done by the field $\mathbf{E}_{\text{rad}}$ on the oscillating dipole. This work represents the transfer of energy stored in the oscillating dipole to the EM field, which then transports this energy to infinity in the form of radiation (spherical waves). Thus, we have

$$P_{\text{scat}} = \oint_{\partial v} \langle \mathbf{S}_{\text{rad}} \rangle \cdot \hat{n} da = \frac{1}{2} \int \text{Re}\Big[\Big(\mathbf{E}_{\text{rad}} \times \mathbf{H}^{*}_{\text{rad}}\Big) \cdot \hat{r}\Big] r^2 d\Omega, \tag{2.55}$$

which leads us to conclude that, in fact, equation (2.41) describes the power scattered (or, alternatively, emitted or radiated) by the dielectric sphere. Thus, the power scattered (P_{scat}) by a dielectric sphere in a medium with dielectric constant ε_{m} is

$$P_{\text{scat}} = \frac{c^2 Z_{\text{m}} k_{\text{m}}^4}{12\pi\varepsilon_{\text{m}}} |\mathbf{p}|^2, \tag{2.56}$$

where $Z_{\text{m}} = \sqrt{\mu_0 \mu_{\text{m}} / \varepsilon_0 \varepsilon_{\text{m}}}$ is the impedance of the medium (for simplicity, let's take $\mu_{\text{m}} = 1$). The scattering cross-section can be written as

$$\sigma_{\text{scat}} = \frac{P_{\text{scat}}}{I} = \frac{k_{\text{m}}^4}{6\pi} |\alpha_{\text{rad}}|^2. \tag{2.57}$$

For wavelengths where $k_{\text{m}} a \ll 1$, we can disregard the correction from equation (2.43) and use the Clausius–Mossotti relation to describe the polarizability (α_{CM}), instead of using α_{rad} directly. Figure 2.8(a) compares the polarizability described by the Clausius–Mossotti relation with that obtained by applying the radiation correction as a function of particle radius for $\lambda = 514$ nm ($k_{\text{m}} = 0.016$ nm^{-1} in water). Note that for $a < 60$ nm, we have $\alpha_{\text{CM}} \approx \text{Re}\{\alpha_{\text{rad}}\}$. For $a > 60$ nm, where $k_{\text{m}} a > 1$, the radiation correction becomes necessary. However, for larger wavelengths, the size range where both polarizabilities yield similar values is broader. For instance, for $\lambda = 1064$ nm, this range extends up to approximately $a = 100$ nm.

Therefore, for particles with a real dielectric constant and $k_{\text{m}} a \ll 1$, the scattering cross-section calculated using α_{CM} will be close to that calculated with α_{rad}, as shown in figure 2.8(b). Similarly, if the dielectric constant is complex and $k_{\text{m}} a \ll 1$, we can replace α_{rad} in equations (2.49) and (2.57) with the Lorentz–Lorenz relation and rewrite the extinction and scattering cross-sections as

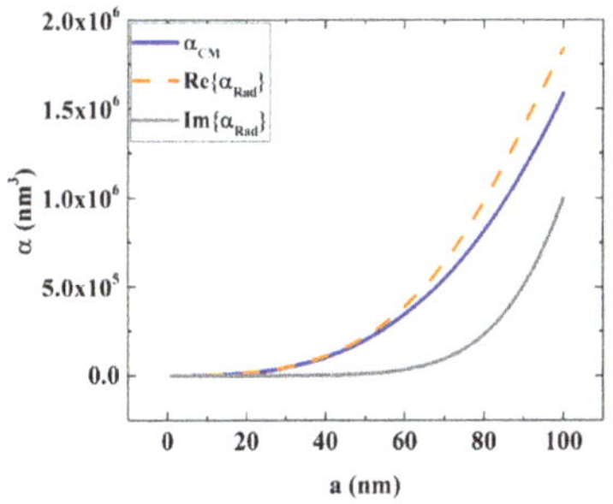

(a) Polarizability as a function of radius.

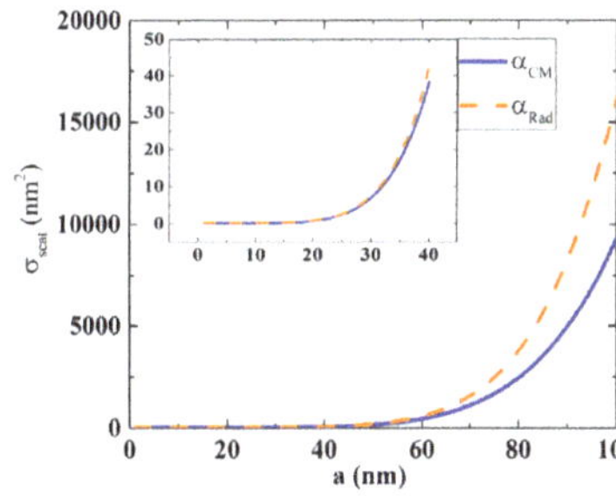

(b) Scattering cross section as a function of radius.

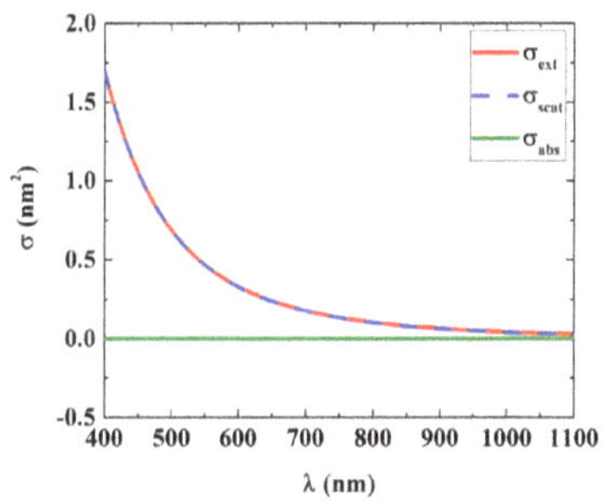

(c) Cross sections in a dielectric particle as a function of wavelength.

Figure 2.8. Polarizability and cross-sections of a dielectric particle. (a) Polarizability of a dielectric particle as a function of the particle radius. Dark blue solid line: polarizability described by the Clausius–Mossotti relation α_{CM} (equation (2.24)). Orange dashed line: real part of the effective polarizability, including radiation reaction α_{rad} (equation (2.43)). Green solid line: imaginary part of α_{rad}. (b) Scattering cross-section of a dielectric particle as a function of the particle radius. Dark blue line: scattering cross-section calculated using α_{CM} (equation (2.59)). Orange dashed line: scattering cross-section calculated using α_{rad} (equation (2.57)). Inset: zoom showing the behavior of the cross-section for $a < 40$ nm. (c) Cross-sections of a dielectric particle as a function of wavelength. Red solid line: extinction cross-section. Blue dashed line: scattering cross-section. Green solid line: absorption cross-section. In all cases, the cross-sections were calculated using α_{rad}. **Parameters used**: (a) and (b): $\lambda = 514$ nm. (c) $a = 20$ nm. All: $\varepsilon_{\text{m}} = 1.77$, $\varepsilon_{\text{p}} = 2.53$.

$$\sigma_{\text{ext}} = 4\pi k_{\text{m}} a^3 \,\text{Im}\left\{\frac{\varepsilon_{\text{p}} - \varepsilon_{\text{m}}}{\varepsilon_{\text{p}} + 2\varepsilon_{\text{m}}}\right\}, \tag{2.58}$$

$$\sigma_{\text{scat}} = \frac{8\pi}{3} k_{\text{m}}^4 a^6 \left|\frac{\varepsilon_{\text{p}} - \varepsilon_{\text{m}}}{\varepsilon_{\text{p}} + 2\varepsilon_{\text{m}}}\right|^2. \tag{2.59}$$

It is important to emphasize that the size range where equations (2.58) and (2.59) remain valid will strongly depend on the value of the complex part of ε_{p}.

We can also define an *absorption cross-section* as being

$$\sigma_{\rm abs} = \sigma_{\rm ext} - \sigma_{\rm scat} = k_{\rm m}\,{\rm Im}\{\alpha_{\rm rad}\} - \frac{k_{\rm m}^4}{6\pi}|\alpha_{\rm rad}|^2. \tag{2.60}$$

Equation (2.60) defines the absorption cross-section ($\sigma_{\rm abs}$) in terms of the extinction cross-section and the scattering cross-section, according to the relation $\sigma_{\rm ext} = \sigma_{\rm sct} + \sigma_{\rm abs}$. Extinction refers to the total energy taken from the incident wave, which is partly re-emitted as radiation (scattering) and partly absorbed by the dielectric sphere, converting to other forms of energy such as heat or by exciting the atoms of the material. Thus, the absorption cross-section can be interpreted as an 'effective area' that quantifies the effectiveness of the dielectric sphere in absorbing the energy of the incident EM wave, representing the probability of converting this energy into other forms.

Note that the intrinsic properties of the sphere, such as its size and the complex nature of its dielectric constant, significantly influence how energy is extinguished from the incident wave. In conventional dielectric spheres, where the dielectric constant is a positive real number, the polarizability is described by $\alpha_{\rm rad}$, and extinction is predominantly by scattering $\sigma_{\rm ext} \approx \sigma_{\rm sct}$ as shown in figure 2.8(c). In contrast, for materials with complex dielectric constants (as can be the case for some dielectrics and more precisely for metals and some semiconductors), the energy absorption by the sphere becomes significant and the absorption cross-section becomes the extinction dominant, leading to the relation $\sigma_{\rm ext} \approx \sigma_{\rm abs}$.

Figure 2.8(c) illustrates the variation of the extinction, scattering, and absorption cross-sections, calculated using $\alpha_{\rm rad}$, for a particle with a real dielectric constant as a function of wavelength. Here we consider for the entire wavelength range the parameters $\varepsilon_{\rm p} = 2.53$, $\varepsilon_{\rm m} = 1.77$, and $a = 20$ nm. From the cross-sections, we can directly determine the scattering, absorption, and extinction powers, knowing the intensity of the incident wave and the properties of the medium and the dielectric sphere. This approach facilitates the calculation of the forces exerted on the sphere, as will be discussed in the next section.

2.2 Optical forces

In EM fields, as in mechanical systems, wave propagation is associated with both energy and momentum transport. The energy flux is described by the Poynting vector, which represents the rate of energy transfer per unit area. Thus, **S** gives us both the direction and intensity of the energy flux of the EM waves.

However, this energy flux also carries with it linear momentum. The linear momentum density of EM fields, represented by $\mathbf{g}_{\rm em}$, is directly related to the energy flux through the following equation:

$$\mathbf{g}_{\rm em} = \frac{\mathbf{S}}{c^2}. \tag{2.61}$$

This result shows that, just as **S** transports wave energy, it also transports linear momentum.

This relationship between energy flux and momentum density reflects a fundamental characteristic of EM radiation: it not only carries energy but also transfers momentum to matter during interactions. Consequently, wave propagation, quantified by the Poynting vector, involves the simultaneous transfer of energy and momentum when radiation interacts with matter.

It is crucial to emphasize that energy and momentum are intrinsic properties of EM fields, regardless of whether the fields propagate through space. Even in static configurations, where there is no propagation, the presence of an energy flux ($\mathbf{S} \neq 0$) implies an associated momentum density. This highlights that energy and momentum arise inherently from the fields themselves and result from the interaction between the electric and magnetic components. At any point where the cross-product of the electric field and the magnetic field is non-zero, there is a momentum density and an energy flux, both described by the Poynting vector.

Evidently, stationary EM fields generate a stationary energy flux, which implies that $\partial_t \mathbf{g}_{em} = 0$. Furthermore, in many situations of interest, the linear momentum density $\mathbf{g}_{em}$ is zero. A classic example is the interaction between two point charges at rest. In this case, the linear momentum density of the fields is zero since there is no relative motion between the charges and the associated fields are purely electrostatic ($\mathbf{H} = \mathbf{0}$).

From this, it might be natural to conclude that, in such situations, there is no force of EM origin, since force, by definition, is related to the temporal variation of linear momentum. However, our everyday experience shows the opposite. We know that a point charge A at rest exerts a force on another point charge B, also at rest, and that this force originates from the electrostatic field generated by charge A. This is the Coulomb force, whose form is well known. The question then arises: how can an electrostatic field, in which $\mathbf{g}_{em} = 0$, exert a force on a material object (the point charge B)?

The answer to this question lies in the fact that, although the linear momentum density of the field, $\mathbf{g}_{em}$ is zero in static fields, the presence of a material object introduces disturbances in the field lines, creating a buildup of EM stresses in space. These stresses can generate forces on the object.

To understand this we can use a useful analogy: an object immersed in a fluid. When an object is placed in a fluid, the fluid exerts stresses (pressures and shear stresses) on the surface of the object. These stresses arise because the object displaces some of the fluid around it, altering the fluid flow lines. Consequently, the fluid exerts a force on the object.

Similarly, in the case of EM fields, the presence of an object (such as a dielectric or conducting sphere) alters the distribution of the fields around it. Figure 2.2(a) illustrates this change, showing the electric field lines around a dielectric sphere immersed in a uniform electric field. Near the surface of the sphere, the initially uniform field lines become distorted, bending around the sphere due to the polarization induced in the material.

This distortion in the field lines occurs because the material of the sphere, being dielectric, becomes polarized under the action of the external electric field, creating internal dipoles aligned with the incident electric field. The additional fields generated by these dipoles add to the incident field, resulting in a new configuration of field lines around the object. This redistribution of the field generates a specific pattern of EM stress, which are responsible for the forces exerted on the object.

EM stresses are described by the Maxwell stress tensor ($\mathbf{T}_\mathrm{M}$), whose ij component is given by

$$T_\mathrm{M}^{ij} = \varepsilon_0\left(E_iE_j - \frac{1}{2}\delta_{ij}|\mathbf{E}|^2\right) + \frac{1}{\mu_0}\left(B_iB_j - \frac{1}{2}\delta_{ij}|\mathbf{B}|^2\right), \tag{2.62}$$

where E_k and B_k ($k = i, j$) are the kth components of the total electric field and total magnetic field, respectively.

Physically, $-T_\mathrm{M}^{ij}$ represents the flux density of the ith component of the linear momentum along the j direction. The role of $\mathbf{T}_\mathrm{M}$ in describing the transport and conservation of linear momentum can be elucidated through the continuity equation, which relates the temporal variation of the linear momentum of the fields to the EM forces in the system,

$$\frac{\partial \mathbf{g}_\mathrm{em}}{\partial t} - \nabla \cdot \mathbf{T}_M = -\frac{\partial \mathbf{g}_\mathrm{mat}}{\partial t}, \tag{2.63}$$

where $\mathbf{g}_\mathrm{mat}$ is the linear momentum density contained in the matter. Equation (2.63) tells us that, if the linear momentum of the fields changes at a point, this occurs due to the transport of momentum, represented by the flux $-\mathbf{T}_\mathrm{M}^{ij}$, or the transfer of momentum from the fields to the matter, generating a force on the latter.

Let v be an arbitrary volume in space, with ∂v as the surface delimiting v. Integrating equation (2.63) over this volume and applying Gauss's theorem for tensors, we obtain

$$\frac{d}{dt}\int_v \mathbf{g}_\mathrm{em} dv + \frac{d}{dt}\int_v \mathbf{g}_\mathrm{mat} dv = \oint_{\partial v} \hat{n} \cdot \mathbf{T}_\mathrm{M} da, \tag{2.64}$$

$$\frac{\partial\left(\mathbf{P}_\mathrm{em}^v + \mathbf{P}_\mathrm{mat}^v\right)}{\partial t} = \oint_{\partial v} \hat{n} \cdot \mathbf{T}_M da, \tag{2.65}$$

where $\mathbf{P}_\mathrm{em}^v$ and $\mathbf{P}_\mathrm{mat}^v$ are the EM and matter linear momenta contained in the volume v, respectively. Equation (2.65) states that the variation of the total momentum inside the volume v, i.e. the momentum transfer between fields and matter, corresponds to the EM momentum flow across the surface delimiting v.

The resultant force that EM fields exert on the matter contained in v can be expressed as

$$\mathbf{F}_\mathrm{mat}^v = -\frac{d}{dt}\int_v \frac{\mathbf{S}}{c^2} dv + \oint_{\partial v} \hat{n} \cdot \mathbf{T}_\mathrm{M} da, \tag{2.66}$$

where $\mathbf{g}_\mathrm{em} = \mathbf{S}/c^2$.

Equation (2.66) provides the total force that EM fields exert on the charge and current densities present in the matter contained in v. However, the resultant force directly acting on these charge and current distributions is known as the Lorentz force, whose expression (per unit volume) is given by

$$\mathbf{f}_{\mathrm{L}} = \rho\mathbf{E} + \mathbf{J} \times \mathbf{B}. \tag{2.67}$$

Thus, $\mathbf{F}_{\mathrm{mat}} = \int_v \mathbf{f}_{\mathrm{L}} dv = \mathbf{F}_{\mathrm{L}}$, where $\mathbf{F}_{\mathrm{L}}$ is the Lorentz force.

The Lorentz force arises from both the temporal variation of the fields' energy flux through the material medium and the stresses exerted by the fields. For example, in electrostatics, the mutual force between two stationary point charges originates solely from the tensor term since $\partial_t \mathbf{S}/c^2 = 0$.

A similar situation occurs for EM waves at optical frequencies. At these frequencies, the field oscillations are much faster than the inertial response of charges and currents in the matter, so only the time-averaged interactions are significant. Thus, at optical frequencies, the force exerted by EM fields on the charge and current densities in the matter contained in v is given by the time average of equation (2.67)

$$\mathbf{F}_{\mathrm{op}} = \langle \mathbf{F}_{\mathrm{L}} \rangle = -\frac{d}{dt}\int_v \frac{\langle \mathbf{S} \rangle}{c^2} dv + \oint_{\partial v} \hat{n} \cdot \langle \mathbf{T}_{\mathrm{M}} \rangle da. \tag{2.68}$$

Since the first term on the right-hand side is zero ($\partial_t \langle \mathbf{S} \rangle = 0$), the optical force originates solely from the stresses exerted by the fields on the matter. Thus, we have:

$$\mathbf{F}_{\mathrm{op}} = \oint_{\partial v} \hat{n} \cdot \langle \mathbf{T}_{\mathrm{M}} \rangle da. \tag{2.69}$$

Equation (2.69) applies to any field and particle configuration and is directly derived from the continuity equation of linear momentum. For harmonic fields in a non-absorbing medium with refractive index n_{m} and dielectric constant ε_{m}, equation (2.69) can be rewritten as

$$\mathbf{F}_{\mathrm{op}} = \frac{1}{2}\varepsilon_{\mathrm{m}} r^2 \mathrm{Re}\left\{ \oint_{\partial v} \hat{r} \cdot \left[\mathbf{E} \otimes \mathbf{E}^* + \frac{c^2}{n_{\mathrm{m}}^2} \mathbf{B} \otimes \mathbf{B}^* - \frac{1}{2}\left(|\mathbf{E}|^2 + \frac{c^2}{n_{\mathrm{m}}^2} |\mathbf{B}|^2 \right) \mathbf{I}_{\mathrm{d}} \right] d\Omega \right\}, \tag{2.70}$$

where $\mathbf{I}_{\mathrm{d}}$ is the identity tensor, $\mathbf{E}$ and $\mathbf{B}$ are the total fields (incident field + scattered field), and ∂v is any surface enclosing the particle of interest.

Although equation (2.70) is very general and correctly describes the optical force on a particle of any size and shape, directly solving equation (2.70) is not always straightforward. In fact, any case more complex than a homogeneous sphere becomes impractical due to the need of knowing the scattering fields on the particle, a classical problem in the OT literature known as 'the scattering problem' [2, 3].

From a practical perspective, it is always advantageous to calculate optical forces by exploring some symmetry or simplification in the problem. For example, if the particle is spherical and has a radius $a \gg \lambda$, the use of geometrical optics to compute optical forces is an excellent approximation, as will be evident in chapter 3. On the other hand, if $a \ll \lambda$, the quasi-static approximation or direct calculation of the force using the Lorentz force averaged in time is an efficient solution, provided the incident electric field can be considered constant inside the particle [4].

For the intermediate regime, where $a \approx \lambda$, there are few alternatives other than to calculate the force directly from the equation (2.70). In the following sections, we will use the Rayleigh approximation to calculate the optical force $\mathbf{F}_{\mathrm{op}}$, while in the next chapter, we discuss the calculation of such force in the context of the geometrical optics regime.

2.2.1 Optical force in the $a \ll \lambda$ limit

As discussed, the optical force arises as a result of the stresses that EM fields exert on the matter when it is illuminated. Since calculating the flow integral of the Maxwell stress tensor (equation (2.69)) requires knowing $\mathbf{T}_{\mathrm{M}}$ at the integration surface, for non-absorbing media it is always possible to extend the integration area to the radiation zone. In this case, the problem essentially becomes determining the scattering field generated by the particle. Once the scattering fields are known, all the work is reduced to adding them to the incident field and substituting the total field into equation (2.70). Unfortunately, this approach is cumbersome even for simple geometries, such as the case of a homogeneous spherical particle.

Therefore, whenever possible, it is advantageous to explore the symmetries or approximations allowed by the problem. Let us now consider the case where the dielectric particle is spherical and its radius (a) is much smaller than the wavelength of the radiation (in vacuum) λ. In this scenario, the particle can be analyzed as an electric dipole, and the optical force can be obtained as the time-averaged force that the incident fields and the scattering fields exert on the induced charges and currents in the particle. In other words, the force on the particle can be calculated from the time-averaged Lorentz force.

In fact, equation (2.68) shows that the optical force can be interpreted as the Lorentz force averaged over time. Since the temporal variation of the fields is much faster than the inertial response of the charges and currents, so that only $\langle \mathbf{S} \rangle$ is significant, the time-averaged Lorentz force will be equivalent to the average of the Maxwell stress tensor flow.

The force on a dipole immersed in a non-absorbing medium, with dielectric constant ε_{m}, can be expressed in terms of the polarizability $\alpha_d = \alpha_{\mathrm{rad}}$ (which reduces to the Lorentz–Lorenz polarizability when $k_{\mathrm{m}} a \ll 1$ and ε_{p} is complex), the extinction cross-section σ_{ext}, the Poynting vector of the incident wave $\mathbf{S}_{\mathrm{i}}$, and the spin density of the incident wave ($\mathbf{s}_d$), which will be defined later. The calculations for obtaining the force on the dipole can be found in reference [2], whose result is

$$\mathbf{F}_{\mathrm{op}} = \frac{1}{4}\varepsilon_0 \varepsilon_{\mathrm{m}} \, \mathrm{Re}(\alpha_d) \boldsymbol{\nabla} |\mathbf{E}_i|^2 + \frac{\sqrt{\varepsilon_{\mathrm{m}}}\,\sigma_{\mathrm{ext}}}{c} \langle \mathbf{S}_{\mathrm{i}} \rangle - \frac{1}{2} \frac{\sigma_{\mathrm{ext}} c}{\sqrt{\varepsilon_{\mathrm{m}}}} \boldsymbol{\nabla} \times \mathbf{s}_d. \tag{2.71}$$

The force on the dipole, given by equation (2.71), indeed becomes the force on the particle, provided that the electric field is constant within the volume of the particle, as assumed in the quasi-static approximation. Information about the particle size is incorporated in the terms of the polarizability α_{d} and, consequently, in the term of the extinction cross-section. Note that all the fields present in equation (2.71) refer to the incident wave. The information about the scattered fields, on the other hand, is incorporated into the extinction cross-section and also into α_d through the self-interaction term.

Each term in equation (2.71) describes a mechanism of interaction between radiation and matter, with each contributing distinctly to the optical force. In the context of OT, these terms reveal how the radiation interacts with (and to generate) the trapped particles. A detailed analysis of each term allows for an understanding of how the resulting force, which manipulates the particles, is distributed among the different interaction mechanisms, such as polarization and scattering. These terms are crucial for understanding the dynamics of optical trapping and the precise control of particles in optical systems.

2.2.2 Optical beam

Before analyzing how each term of the optical force acts on a dielectric particle, it is essential to describe the fields of the incident wave, i.e. to determine the electric field vector $\mathbf{E}_i$ and the magnetic field vector $\mathbf{H}_i$ of the beam illuminating the particle. Specifically, we are interested in the optical force generated by focusing a beam through an objective lens with numerical aperture NA, which corresponds to studying the optical forces in an optical trap.

Most OT operate in the TEM_{00} mode, where the electric and magnetic fields are mutually orthogonal and perpendicular to the beam propagation direction, assumed to be along z. This mode is a solution of the Helmholtz equation in the paraxial approximation[5], resulting in a Gaussian amplitude profile in the transverse plane. In this regime, the beam is characterized by a single parameter: the beam waist $w_0 = \lambda/2\pi n_m NA$, simplifying the field description to scalar equations.

The adopted geometry for this chapter is illustrated in figure 2.9. We consider a dielectric sphere of radius a and ε_p immersed in a medium ε_m, illuminated by a fundamental Gaussian beam, linearly polarized along $\hat{x}$. The coordinate system is centered at O_G, the beam waist, while the centroid of the particle is positioned at $\mathbf{r}_d = (x, y, z)$.

Within the paraxial approximation, the incident electric field at position $\mathbf{r}$ is given by [4]

$$\mathbf{E}_i(\mathbf{r}) = E_0(\mathbf{r})\hat{x} = E_0 \frac{ikw_0^2}{ikw_0^2 + 2z} \exp(-ikz) \exp\left[-i\frac{2kz(x^2 + y^2)}{(kw_0^2)^2 + (2z)^2}\right] \exp\left[-\frac{(kw_0)^2(x^2 + y^2)}{(kw_0^2)^2 + (2z)^2}\right]\hat{x}, \quad (2.72)$$

where k is the wave number in the medium and E_0 is the field amplitude at the beam center. The corresponding magnetic field is given by

[5] In the paraxial approximation, beam propagation occurs predominantly along the axial direction z, with small transverse variations. This simplifies Maxwell's equations, making the mathematical analysis more tractable. This approximation is valid for beams that are not strongly focused, for which propagation angles are small.

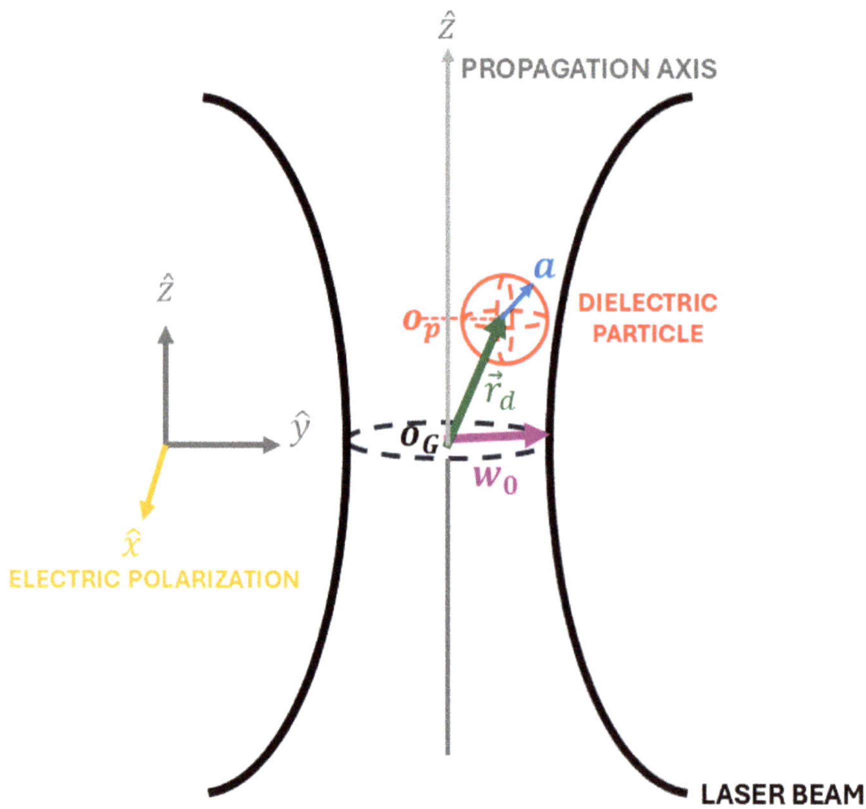

Figure 2.9. Diagram illustrating the geometry used in this section.

$$\mathbf{H}_i(\mathbf{r}) = \hat{z} \times \frac{\mathbf{E}_i(\mathbf{r})}{Z_\mathrm{m}} = n_\mathrm{m}\varepsilon_0 c\mathbf{E}_i(\mathbf{r})\hat{y}, \tag{2.73}$$

where Z_m is the impedance of the medium. The real physical fields are time- and space-dependent functions[6]:

$$\mathcal{E}_i(\mathbf{r}, t) = \mathrm{Re}[\mathbf{E}_i(\mathbf{r})exp(-i\omega t)] \tag{2.74}$$

$$\mathcal{H}_i(\mathbf{r}, t) = \mathrm{Re}[\mathbf{H}_i(\mathbf{r})exp(-i\omega t)] \tag{2.75}$$

The energy flux density of the EM wave is described by the Poynting vector,

$$\mathcal{S}_i(\mathbf{r}, t) = \mathcal{E}_i(\mathbf{r}, t) \times \mathcal{H}_i(\mathbf{r}, t) \tag{2.76}$$

The beam irradiance is obtained as the time average of the Poynting vector,

$$\mathbf{I}(\mathbf{r}) = \langle \mathcal{S}_i(\mathbf{r}, t) \rangle = \frac{1}{2}\mathrm{Re}\left[\mathbf{E}_i(\mathbf{r}) \times \mathbf{H}_i^*(\mathbf{r})\right] = \frac{n_\mathrm{m}\varepsilon_0 c}{2}|\mathbf{E}_i(\mathbf{r})|^2\hat{z} = I(\mathbf{r})\hat{z}, \tag{2.77}$$

[6] Here we modify the way of writing the real part of the fields ($\mathcal{E}$, $\mathcal{H}$) to avoid confusion with the term that describes the spatial dependence of the fields in complex notation (**E**, **H**).

where

$$I(\mathbf{r}) = \left(\frac{2P}{\pi w_0^2}\right)\frac{1}{1+(2\tilde{z})^2}\exp\left[-\frac{2(\tilde{x}^2+\tilde{y}^2)}{1+(2\tilde{z})^2}\right]. \tag{2.78}$$

Here, P represents the beam power, while $(\tilde{x}, \tilde{y}, \tilde{z})$ are normalized spatial coordinates defined as $(\tilde{x}, \tilde{y}, \tilde{z}) = (x/w_0, y/w_0, z/kw_0^2)$.

It is important to note that the above equations do not accurately describe strongly focused Gaussian beams, as these cannot be treated under the paraxial approximation. A relevant parameter in this context is the dimensionless factor s,

$$s = \frac{1}{kw_0} = \frac{\lambda}{2\pi n_{\mathrm{m}} w_0}. \tag{2.79}$$

When s becomes significant, non-paraxial effects must be taken into account, requiring more sophisticated descriptions, such as the corrections proposed by Davis [5] and Barton and Alexander [6, 7]. According to Barton's work, the error in describing the fields $\mathbf{E}_i$ and $\mathbf{H}_i$ using the zeroth-order Gaussian approximation, compared to their description with a fifth-order correction, is approximately 0.817% for $s = 0.02$ (corresponding to $w_0 = 8\lambda/n_{\mathrm{m}}$) and 4.37% for $s = 0.1$ (corresponding to $w_0 = 1.6\lambda/n_{\mathrm{m}}$).

It is important to keep in mind that the field descriptions given by equations (2.72) and (2.73) are merely paraxial approximations of the scalar wave equation for a Gaussian beam and correspond to a zeroth-order approximation in s. Thus, as long as $s \ll 1$, meaning $\lambda \ll w_0$, the description of the fields in terms of a single transverse component remains highly accurate.

This condition imposes an important constraint on the optical forces, particularly on the extinction force and the spin–curl force. Since $\mathbf{S}_{\mathrm{i}}$, has only a component along the propagation axis, these forces do not have transverse components. In other words, in the paraxial approximation, only the gradient force has a transverse component.

At the end of this chapter, we will discuss the work of Harada and Asakura [4], who graphically compare the behavior of optical forces predicted in the Rayleigh regime using the paraxial approximation with that obtained from the generalized Lorenz–Mie theory through a full vectorial description of the fields. We will see that in many cases, the paraxial approximation provides quite satisfactory results.

2.2.3 Gradient force

The first term of equation (2.71) is the main contributor to the formation of optical traps in conventional single-beam OT. This term is referred to as the **gradient force** since it is directly related to the spatial variation of the intensity of the incident light electric field. The origin of this force lies in the electrostatic interaction between the electric field of the wave and the charges induced on the particle surface. Thus, it is a **conservative force**, fundamentally Coulombian in nature.

To understand how this force arises, let us consider a dielectric particle in the quasi-static approximation. In this regime, the electric field can be treated as approximately uniform over the particle's volume, leading to its polarization. This polarization results in the formation of a surface charge distribution and the particle behaves as a dipole whose moment is given by equation (2.22).

Without loss of generality, let us assume that the entire positive surface charge density is $+q$, located at position $\mathbf{r}_+ = \mathbf{r}_d + \delta\mathbf{r}/2$. Similarly, let us assume that the entire negative surface charge density is $-q$, located at position $\mathbf{r}_- = \mathbf{r}_d - \delta\mathbf{r}/2$. Here, $\mathbf{r}_d$ is the centroid position of the particle, and $\delta\mathbf{r} = \mathbf{r}_+ - \mathbf{r}_-$ is the separation between the charges.

In addition to inducing polarization, the electric field also exerts forces on the induced charges. Let $\mathbf{F}_\pm(\mathbf{r}_\pm)$ be the forces that $\mathbf{E}_i$ exerts on $\pm q$ at positions $\mathbf{r}_\pm$. Since the interaction is electrostatic, the resulting force $\mathbf{F}_G$ acting on the centroid of the particle is given by

$$\mathbf{F}_G(\mathbf{r}_d) = \mathbf{F}_+(\mathbf{r}_+) + \mathbf{F}_-(\mathbf{r}_-) = q[\mathbf{E}_i(\mathbf{r}_+) - \mathbf{E}_i(\mathbf{r}_-)]. \tag{2.80}$$

It becomes evident that if $\mathbf{E}_i$ is uniform throughout space, then $\mathbf{F}_G = 0$. However, if $\mathbf{E}_i$ is not uniform, such that $\nabla\mathbf{E}_i \neq 0$, we have $\mathbf{F}_G \neq 0$. To clarify this further, we expand $\mathbf{E}_i(\mathbf{r}_\pm)$ in a Taylor series around $\mathbf{r}_d$,

$$\mathbf{E}_i(\mathbf{r}_\pm) \approx \mathbf{E}_i(\mathbf{r}_d) \pm (\delta\mathbf{r} \cdot \nabla)\mathbf{E}_i(\mathbf{r}_d). \tag{2.81}$$

Substituting 2.81 into 2.80, we obtain

$$\mathbf{F}_G(\mathbf{r}_d) = q(\delta\mathbf{r} \cdot \nabla)\mathbf{E}_i(\mathbf{r}_d) = (\mathbf{p}(\mathbf{r}_d) \cdot \nabla)\mathbf{E}_i(\mathbf{r}_d) \tag{2.82}$$

The expression $(\mathbf{p}(\mathbf{r}_d) \cdot \nabla)\mathbf{E}_i(\mathbf{r}_d)$ represents the instantaneous force on the dipole. However, in optical fields, the quantity of interest is the time-averaged force, $\langle\mathbf{F}_G\rangle$, since the particle's inertial response is much slower than the light's frequency. To find this average, we must consider that the dipole moment $\mathbf{p}$ is itself induced by the field $\mathbf{E}_i$ and oscillates with it. The rigorous time-averaging of the instantaneous force, taking into account the relation $\mathbf{p} = \epsilon_0\epsilon_m\alpha_d\mathbf{E}_i$, yields the final expression for the gradient force:

$$\mathbf{F}_G(\mathbf{r}_d) = \frac{1}{4}\varepsilon_0\varepsilon_\mathrm{m}\,\mathrm{Re}\{\alpha_d\}\,\nabla\,|\mathbf{E}_i(\mathbf{r}_d)|^2, \tag{2.83}$$

from which we obtain the first term of equation (2.71).

The gradient force is directly related to $\mathrm{Re}\{\alpha_d\}$, meaning that its direction depends on the arrangement of the polarization charges on the particle surface. In other words, $\mathbf{F}_G$ can either attract or repel the particle, depending on the relationship between the dielectric constant of the particle (ε_p) and that of the medium (ε_m).

This can be visualized in figure 2.10, which illustrates the charge distribution in two distinct cases: in figure 2.10(a), we have $\varepsilon_\mathrm{p} > \varepsilon_\mathrm{m}$, while in figure 2.10(b), we have $\varepsilon_\mathrm{p} < \varepsilon_\mathrm{m}$. In both cases, the electric field intensity $|\mathbf{E}\,|^2$ follows a Gaussian profile.

When $\varepsilon_\mathrm{p} > \varepsilon_\mathrm{m}$, the dipole moment $\mathbf{p}$ is parallel to $\mathbf{E}_i$. Thus, the positive charges accumulate closer to the region of higher field intensity. As a result, the attractive force on the positive charges is higher than the repulsive force on the negative

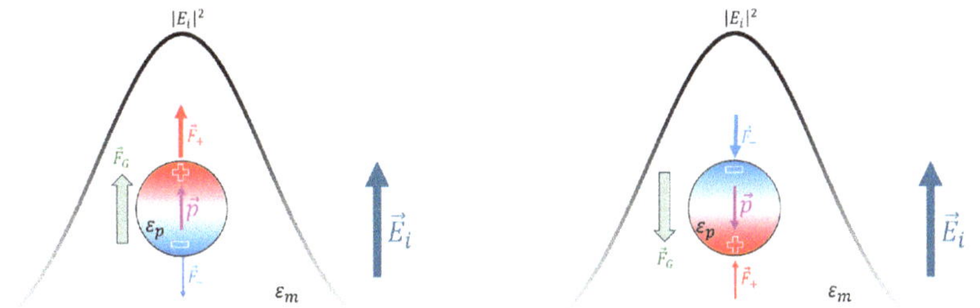

(a) Coulomb forces for $\varepsilon_p > \varepsilon_m$. (b) Coulomb forces for $\varepsilon_p < \varepsilon_m$.

Figure 2.10. Distribution of Coulomb forces generated by the induced charges on the particle.

charges, meaning that $|\mathbf{F}_+| > |\mathbf{F}_-|$. Consequently, the particle is driven toward the region of higher field intensity. In this configuration, the **gradient force is attractive**.

On the other hand, when $\varepsilon_\text{p} < \varepsilon_\text{m}$, the dipole moment $\mathbf{p}$ is antiparallel to $\mathbf{E}_i$, reversing the charge distribution on the particle surface. In this case, the negative charges are closer to the region of higher field intensity, resulting in $|\mathbf{F}_-| > |\mathbf{F}_+|$, which pushes the particle away from the region of higher field intensity. Thus, the **gradient force is repulsive**.

It is important to note that the interaction between the induced charges and the external field is crucial in determining the behavior of the particle within a non-homogeneous optical field. Depending on the relative permittivity between the particle and the surrounding medium, the gradient force can act as a 'trap' capturing the particle and keeping it near the region of highest field intensity, or as a barrier that prevents it from remaining in high-intensity regions.

The reader may find it contradictory that we assume the quasi-static approximation while stating that the gradient force only exists if the field is non-uniform. The apparent contradiction between the existence of spatial gradients ($\nabla \mathbf{E}_i \neq 0$) and the validity of the quasi-static approximation can be resolved by analyzing the length scales involved. The quasi-static approximation requires the field to be approximately uniform within the particle's volume, which does not prohibit variations on larger scales. The spatial variation of the field occurs over a length scale $L \sim |\mathbf{E}_i|/|\nabla \mathbf{E}_i|$, where, for Gaussian optical beams, $L \approx w_0$. Therefore, for the quasi-static approximation to remain valid in a non-uniform field, it is necessary that $a \ll \lambda$ and $a \ll w_0$. Note that within the zeroth-order Gaussian approximation, where $w_0 > \lambda$, it is sufficient that $a \ll w_0$ holds[7].

Using the irradiance defined in equation (2.77), $|\mathbf{E}_i|^2$ can be written as

$$|\mathbf{E}_i(\mathbf{r}_d)|^2 = \frac{2}{n_\text{m}\varepsilon_0 c} I(\mathbf{r}_d), \tag{2.84}$$

where $I(\mathbf{r}_d)$ is given by equation (2.78).

[7] As we will see when discussing the results of Harada's work [4], this condition can be relaxed to $a \leqslant w_0$ for the gradient force.

Using the geometry from figure 2.9 in Cartesian coordinates and substituting 2.84 into 2.83, we obtain the Cartesian components of the gradient force[8]:

$$\mathbf{F}_{\mathrm{G}x}(\mathbf{r}_d) = -\frac{n_{\mathrm{m}}}{2c}\,\mathrm{Re}\{\alpha_d\}\frac{4\tilde{x}/w_0}{1+(2\tilde{z})^2}\left(\frac{2P}{\pi w_0^2}\right)\frac{1}{1+(2\tilde{z})^2}\exp\left[-\frac{2(\tilde{x}^2+\tilde{y}^2)}{1+(2\tilde{z})^2}\right]\hat{x}, \quad (2.85)$$

$$\mathbf{F}_{\mathrm{G}y}(\mathbf{r}_d) = -\frac{n_{\mathrm{m}}}{2c}\,\mathrm{Re}\{\alpha_d\}\frac{4\tilde{y}/w_0}{1+(2\tilde{z})^2}\left(\frac{2P}{\pi w_0^2}\right)\frac{1}{1+(2\tilde{z})^2}\exp\left[-\frac{2(\tilde{x}^2+\tilde{y}^2)}{1+(2\tilde{z})^2}\right]\hat{y}, \quad (2.86)$$

$$\mathbf{F}_{\mathrm{G}z}(\mathbf{r}_d) = -\frac{n_{\mathrm{m}}}{2c}\,\mathrm{Re}\{\alpha_d\}\frac{8\tilde{z}/(kw_0^2)}{1+(2\tilde{z})^2}\left[1-\frac{2(\tilde{x}^2+\tilde{y}^2)}{1+(2\tilde{z})^2}\right]\left(\frac{2P}{\pi w_0^2}\right)\frac{1}{1+(2\tilde{z})^2}\exp\left[-\frac{2(\tilde{x}^2+\tilde{y}^2)}{1+(2\tilde{z})^2}\right]\hat{z}. \quad (2.87)$$

Note that the gradient force has components in directions where the electric field does not have direct components (the electric field has only the x-component, corresponding to the polarization direction). Another interesting point is that, within the zeroth-order Gaussian approximation, the polarization direction does not affect the gradient force, meaning that $\mathbf{F}_{\mathrm{G}x}(x, 0, 0) = \mathbf{F}_{\mathrm{G}y}(0, y, 0)$. In other words, the optical potential generated by the transverse component of the gradient force is symmetric in the xy plane. The existence of force components in directions where there is no field, as well as the symmetry of the transverse components, arises from the fact that $\mathbf{F}_{\mathrm{G}}$ depends on the gradient of the squared modulus of the electric field, rather than on the gradient of the electric field itself.

Why does $\mathbf{F}_{\mathrm{G}}$ depend on $\nabla|\mathbf{E}_i|^2$ rather than on $\nabla\mathbf{E}_i$, even though its origin lies in the electrostatic interaction of charges with the field? This happens because the polarization charges are proportional to $\mathbf{E}_i$, and the force on these charges is proportional to the field. However, as we discussed earlier, a net force will only exist if the squared modulus of the field varies in space. Therefore, the force depends on the gradient of the squared modulus of the field.

If the paraxial approximation is not valid, for instance, when focusing the beam with a high numerical aperture objective, $|\mathbf{E}_i|^2$ is no longer symmetric in the xy plane, breaking the symmetry of the optical potential as well. The gradient force is affected by the broken symmetry of the beam intensity, making the optical potential polarization-dependent, as extensively discussed in reference [8].

But what exactly is this optical potential? Since the gradient force is conservative, it can be described by a scalar potential U_{opt}, called the optical potential. This potential describes the potential energy that the particle experiences when interacting with the electric field of the incident wave. For particles with positive polarizability, the optical potential is minimized in regions of higher electric field intensity, attracting the particle to these regions. Conversely, for particles with negative polarizability, the optical potential is maximized in these regions, resulting in the repulsion of the particle toward regions of lower field intensity. Using equation (2.84) in equation (2.83), we write

---

[8] Extracted from reference [4].

$$U_{\text{opt}}(\mathbf{r}_d) = -\frac{1}{2}\frac{n_\text{m}}{c}\text{Re}\{\alpha_d\}I(\mathbf{r}_d), \tag{2.88}$$

from which we obtain $\mathbf{F}_\text{G}(\mathbf{r}_d) = -\nabla U_{\text{opt}}(\mathbf{r}_d)$.

Figure 2.11 shows the behavior of the transverse component $F_{\text{G}x}$ of the gradient force as a function of the normalized transverse position (x/w_0) for a Gaussian beam with a waist of $w_0 = 5\ \mu\text{m}$ and $\lambda = 514$ nm ($s = 0.012$, corresponding to an error of less than 1% in describing the EM fields within the zero-order Gaussian approximation [4]). The maximum value of $|\mathbf{F}_G|$ occurs at $(\pm w_0/2, 0, 0)$, $(0, \pm w_0/2, 0)$, and $(0, 0, \pm k w_0^2/2\sqrt{3})$ for the x, y, and z components, respectively. This maximum value depends on both the characteristics of the particle and the optical beam. For a typical polystyrene particle ($\varepsilon_\text{p} = 2.52$) in water ($\varepsilon_\text{m} = 1.77$), the maximum value of the x-component of the gradient force ranges from 3×10^{-7} pN for $a = 5$ nm to 5×10^{-4} pN for $a = 60$ nm.

Figure 2.11(a) illustrates the behavior of $\mathbf{F}_{\text{G}x}$ for cases where $\varepsilon_\text{p} < \varepsilon_\text{m}$. In this situation, where $\mathbf{F}_G$ is repulsive, the particle is directed toward the region of lower field intensity. If the particle is positioned to the right of the optical axis ($x > 0$), $F_{\text{G}x} > 0$, pushing the particle further to the right. The opposite occurs if the particle is to the left of the optical axis. $\mathbf{F}_\text{G}$ remains conservative, meaning that the work required to move the particle is independent of the path taken. However, the potential associated with the gradient force in this scenario is a 'barrier-type' potential, as illustrated on the left side of figure 2.11(a). In this configuration, trapping cannot occur, and $\mathbf{F}_{\text{G}x}$ acts to accelerate the particle.

In figure 2.11(b), we present the behavior of $\mathbf{F}_{\text{G}x}$ and the corresponding optical potential (on the right in the figure) for the case where $F_{\text{G}x}$ is generated in a dielectric particle with $\varepsilon_\text{p} = 2.53$ and $\varepsilon_\text{m} = 1.77$. The arrows indicate the direction of $F_{\text{G}x}$ relative to the centroid position of the particle (x_d). When $x_d > 0$, $F_{\text{G}x} < 0$, pulling the particle to the left. Conversely, when $x_d < 0$, $F_{\text{G}x} > 0$, pushing the particle to the right. This implies that the particle will always be directed toward the optical axis $x = 0$, the region where the beam intensity is maximum and the optical potential is minimal.

In other words, the gradient force acts as a restoring force, and the corresponding optical potential is a *confining optical potential*. In this configuration, the particle can be trapped at $x = 0$. For small displacements around $x = 0$, the optical potential can be approximated by a harmonic potential (dotted blue line), and $F_{\text{G}x}$ can be treated as a Hookean force concerning displacement:

$$U_{opt,x} \approx \frac{1}{2}\kappa_x x^2 \tag{2.89}$$

and

$$F_{\text{G}x} \approx -\kappa_x x. \tag{2.90}$$

In general, we define κ_i, $i = \{x, y, z\}$ as

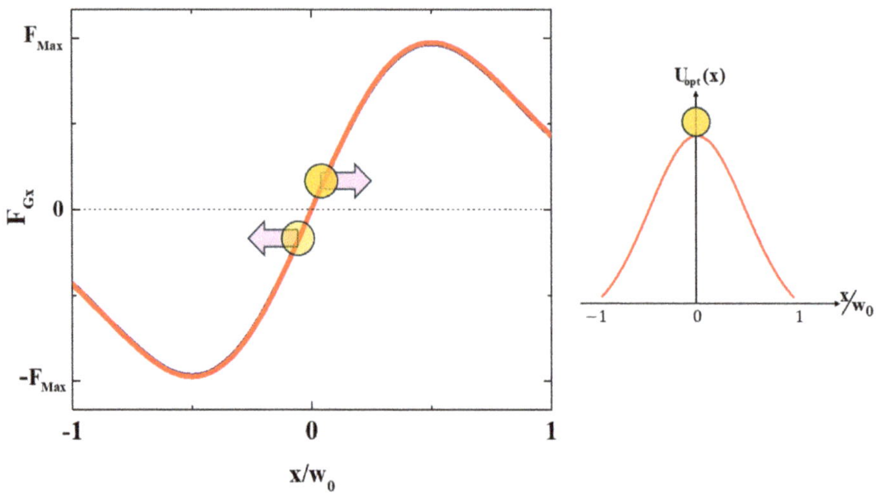

(a) Transverse gradient force and optical potential for $\varepsilon_p < \varepsilon_m$.

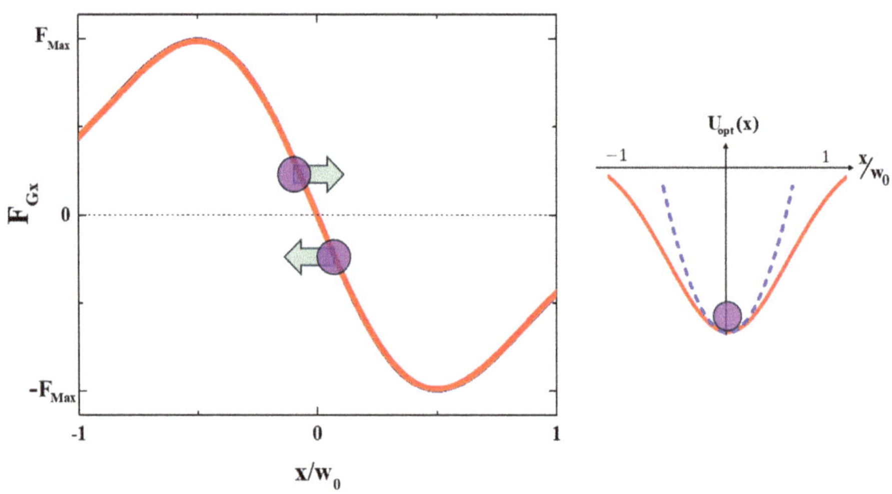

(b) Transverse gradient force and optical potential for $\varepsilon_p > \varepsilon_m$.

Figure 2.11. Behavior of the transverse gradient force ($\mathbf{F}_{Gx}$) and the optical potential as a function of the normalized transverse position (x/w_0) for a Gaussian beam with beam waist $w_0 = 5$ μm and $\lambda = 514$ nm. (a) Case where $\varepsilon_\mathrm{p} < \varepsilon_\mathrm{m}$, showing that the gradient force is repulsive, directing the particle toward the region of lower field intensity. On the right, the optical potential behaves as a barrier. (b) Case where $\varepsilon_\mathrm{p} > \varepsilon_\mathrm{m}$, showing that the gradient force is attractive, directing the particle toward the region of higher field intensity. On the right, the optical potential is confining; the dashed blue line represents the harmonic approximation for small displacements. The arrows indicate the direction of the gradient force relative to the centroid position of the particle (x_d).

$$\kappa_i = -\left.\frac{\partial F_{G_i}}{\partial x_i}\right|_{x_i=0}, \tag{2.91}$$

where $x_i = \{x, y, z\}$. κ_i is the proportionality constant (elastic constant) of the ith component of the gradient force concerning the displacement x_i.

The constant κ describes the strength of the restoring force acting on a particle in an optical trap for a given fixed bead displacement relative to the equilibrium position. In other words, κ defines the stiffness of the trap: the higher κ, the more confined the particle is (higher confining potential). For beams described by the zero-order Gaussian field, $\kappa_x = \kappa_y = \kappa_T$, reflecting the symmetry of the transverse components of F_G. Furthermore, $\kappa_T > \kappa_z$, where κ_z is the longitudinal constant. Since the origin of κ_i is associated with the potential energy the particle experiences when illuminated by the beam, both the beam properties and the particle properties influence the formation of the trap, as measured by κ_i.

In figure 2.12 we present the variation of the transverse trap stiffness κ_T and longitudinal stiffness κ_z as a function of some parameters of interest[9].

Specifically, in figure 2.12(a), we compare κ_T as a function of the radius, calculated from the Clausius–Mossotti polarizability (solid dark blue line) and with the radiation correction (dashed orange line), for a polystyrene particle ($\varepsilon_{\mathrm{p}} = 2.52$) in water ($\varepsilon_{\mathrm{m}} = 1.77$) illuminated by a Gaussian beam with a waist of $w_0 = 5\ \mu$m and $\lambda = 514$ nm. Note that for $a < 45$ nm, the correction to the polarization is negligible, but for $a > 45$ nm, the self-interaction term starts to become relevant, and describing the polarizability using α_{rad} is recommended.

In figure 2.12(b), we analyze how the dielectric properties of the particle influence the formation of the trap by observing the variation of κ_i with the dielectric constant of the particle. The laser characteristics are identical to those in figure 2.12(a). We can observe that κ_i increases monotonically with ε_{p} and, moreover, is negative for $\varepsilon_{\mathrm{p}} < \varepsilon_{\mathrm{m}}$. The fact that $\kappa_i < 0$ implies that the generated optical potential is not confining, meaning the gradient force directs the particle toward the region of lower beam intensity.

It is noticeable that the higher the magnitude of $|\varepsilon_{\mathrm{p}} - \varepsilon_{\mathrm{m}}|$, i.e. the farther ε_{p} is from ε_{m}, the larger $|\kappa_i|$. The constant κ_i is positive if $\varepsilon_{\mathrm{p}} > \varepsilon_{\mathrm{m}}$ (blue region) and negative if $\varepsilon_{\mathrm{p}} < \varepsilon_{\mathrm{m}}$ (red region). This increase in $|\kappa_i|$ is directly related to the increase in the magnitude of the induced charge, as can be seen in equation (2.21). This relation strongly suggests that the increase in polarization charges can be used to construct stronger traps. Indeed, as we will discuss in chapter 4, the resonance between polarization charges and the incident wave electric field in conductive particles is the key to understanding the existence of confinement potentials for these particles and why they are more intense than for dielectric particles. Figure 2.12(c) illustrates the behavior of κ_i as a function of laser power. Here, P represents the

[9] Note that we multiply the longitudinal trap stiffness by a factor of 10^3 for better visualization. All values of κ_i were calculated at the origin of the system, i.e. at position O_G in figure 2.9.

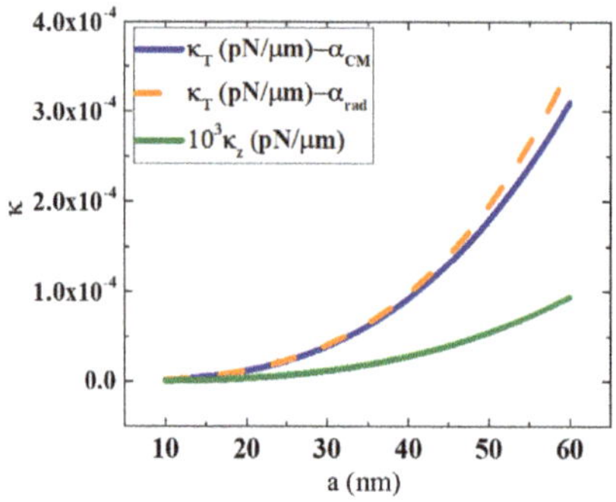

(a) Trap stiffness as a function of particle radius.

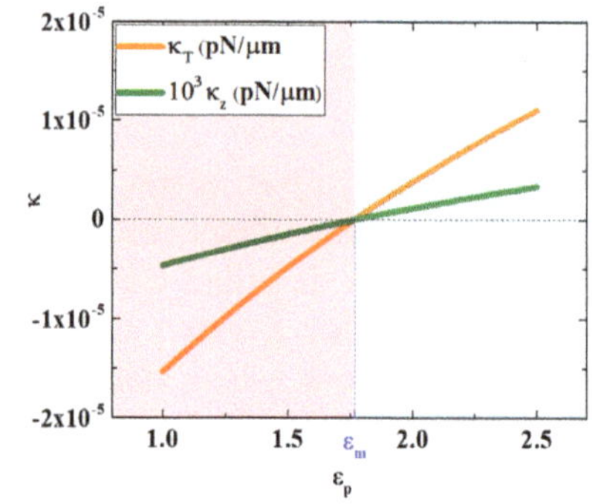

(b) Trap stiffness as a function of the particle dielectric constant.

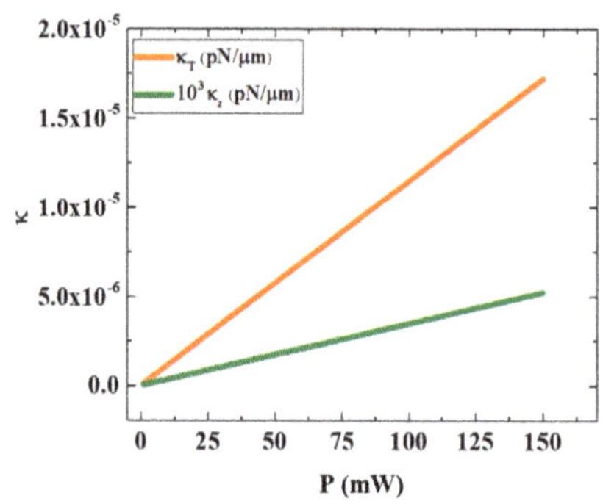

(c) Trap stiffness as a function of the laser beam power.

Figure 2.12. Trap stiffness as a function of various parameters of interest. (a) Variation of κ as a function of the particle radius. Solid blue line: transverse trap stiffness calculated from the polarizability α_{CM}. Dashed orange line: transverse trap stiffness calculated from the polarizability α_{rad}. Solid green line: longitudinal trap stiffness calculated from the polarizability α_{rad} multiplied by 10^3. (b) Variation of κ as a function of the particle dielectric constant calculated from α_{rad}. Solid orange line: transverse trap stiffness. Solid green line: longitudinal trap stiffness multiplied by 10^3. (c) Trap stiffness as a function of the beam power. Solid orange line: transverse trap stiffness. Solid green line: longitudinal trap stiffness multiplied by 10^3. **Parameters used:** (a) $P = 100$ mW, $\varepsilon_{\mathrm{p}} = 2.52$. (b) $P = 100$ mW, $a = 20$ nm. (c) $a = 20$ nm, $\varepsilon_{\mathrm{p}} = 2.52$. In all cases, $w_0 = 5\ \mu$m, $\lambda = 514$ nm, $\mathbf{r}_d = (0, 0, 0)$.

effective power available to interact with the particle, disregarding losses along the optical path such as in lenses, sample holders, objectives, among others.

Although the linear behavior between κ_i and P may seem trivial, given that $\mathbf{F}_G$ depends linearly on power, this result has significant physical meaning: *the interaction between the beam and the particle is not altering the optical properties of the material.* In other words, if the interaction with the beam induced excitations in the particle, such as the generation of charge carriers [9, 10], excitons [11], or other effects [12, 13], this relationship would no longer be linear, as the dielectric constant of the particle would become power-dependent. In chapter 6, we will see how excitation in semiconductor particles modifies their dielectric properties, breaking the linear regime between κ_i and P. An experimental result evidencing this fact will be presented in chapter 7, figure 7.7.

2.2.4 Longitudinal forces: scattering, extinction and gradient forces

The second term in equation (2.71) is the *extinction force* $\mathbf{F}_{\text{ext}}$, also known as the *radiation pressure force*. The extinction force is the total force experienced by a particle due to its interaction with an EM field, accounting for both scattering and absorption processes. It represents the total rate of momentum removed from the EM field by the particle. Just as the extinction cross-section is the sum of the scattering and absorption cross-sections, we have that

$$\mathbf{F}_{\text{ext}} = \frac{n_{\text{m}}}{c}\sigma_{\text{ext}}\langle\mathbf{S}_{\text{i}}(\mathbf{r}_d, t)\rangle = \frac{n_{\text{m}}}{c}\sigma_{\text{scat}}\langle\mathbf{S}_{\text{i}}(\mathbf{r}_d, t)\rangle + \frac{n_{\text{m}}}{c}\sigma_{\text{abs}}\langle\mathbf{S}_{\text{i}}(\mathbf{r}_d, t)\rangle = \mathbf{F}_{\text{scat}} + \mathbf{F}_{\text{abs}}, \quad (2.92)$$

where $\langle\mathbf{S}_{\text{i}}\rangle$ is the time-averaged Poynting vector of the incident field, evaluated at the particle centroid.

The extinction force (and, consequently, the scattering and absorption forces) is a non-conservative force whose direction is determined by the Poynting vector, meaning it points in the direction of beam propagation. In the zeroth-order Gaussian approximation, $\mathbf{F}_{\text{ext}}$, $\mathbf{F}_{\text{scat}}$, and $\mathbf{F}_{\text{abs}}$ have only a longitudinal component. In other words, only the gradient force acts on the particle in the transverse direction under such a condition.

It is important to note that this statement ceases to be valid if the Gaussian fields of the incident wave cannot be described by the zeroth-order approximation ($s \geqslant 1$), which frequently occurs for strongly focused Gaussian beams. In such cases, the force is still described by equation (2.92), but $\langle\mathbf{S}_{\text{i}}\rangle$ acquires components in the transverse direction [14].

For dielectric particles where ε_{p} is real, we have $\sigma_{\text{ext}} \approx \sigma_{\text{scat}}$, which means that the radiation pressure is dominated by the scattering term ($\mathbf{F}_{\text{ext}} \approx \mathbf{F}_{\text{scat}}$). For this reason, it is very common to find authors defining $\mathbf{F}_{\text{ext}}$ as the scattering force, although this identification is only strictly correct for non-absorbing dielectric particles.

For the rest of this chapter we will assume that the dielectric constant of the particle is a real number, so that we can treat the radiation pressure solely in terms of the scattering force. In chapters 4 and 6, we will analyze the effect of the absorption force on radiation pressure in the context of plasmonic and semiconductor particles, where we will treat ε_{p} as a complex number. Thus, by describing σ_{scat} using equation (2.59), we can express the scattering force in terms of the normalized positions

$$\mathbf{F}_{\text{scat}}(\mathbf{r}_d) = \frac{8n_{\text{m}}\pi}{3c}k_m^4 a^6 \left|\frac{\epsilon_p - \epsilon_m}{\epsilon_p + 2\epsilon_m}\right|^2 \left(\frac{2P}{\pi w_0^2}\right)\frac{1}{1 + (2\tilde{z})^2}\exp\left[-\frac{2(\tilde{x}^2 + \tilde{y}^2)}{1 + (2\tilde{z})^2}\right]\hat{z}. (2.93)$$

Figure 2.13 shows the forces acting in the longitudinal direction on a dielectric particle, namely, the longitudinal component of the gradient force (solid red line) and the scattering force (solid blue line). For reasons that will become clear later, we

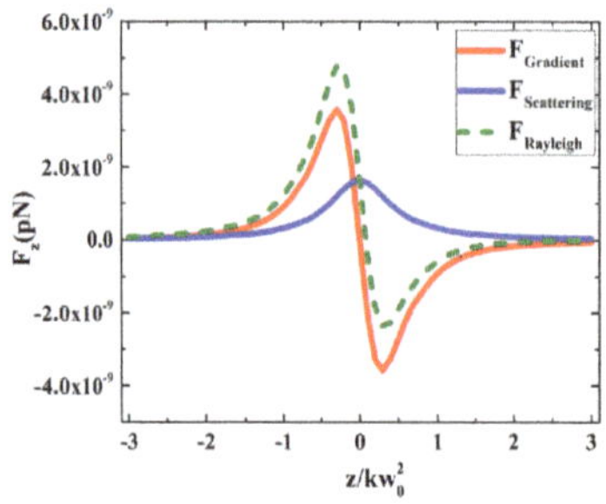

(a) Longitudinal components for $a = 5$ nm and $\varepsilon_p > \varepsilon_m$.

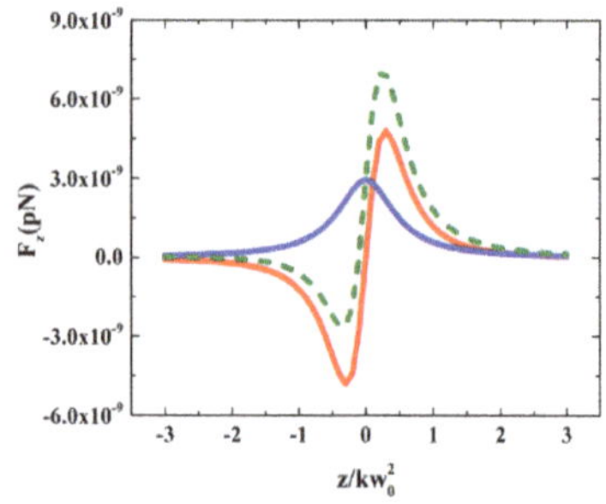

(b) Longitudinal components for $a = 5$ nm and $\varepsilon_p < \varepsilon_m$.

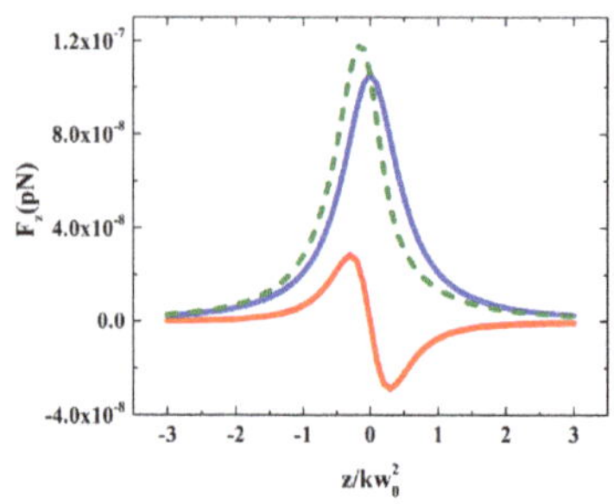

(c) Longitudinal components for $a = 10$ nm and $\varepsilon_p > \varepsilon_m$.

Figure 2.13. Longitudinal components of the forces acting on a dielectric particle. (a) Longitudinal components for $a = 5$ nm and $\varepsilon_{\mathrm{p}} > \varepsilon_{\mathrm{m}}$: The figure shows the longitudinal components of the gradient force (solid red line) and the scattering force (solid blue line) for a polystyrene particle with a radius of 5 nm ($\varepsilon_{\mathrm{p}} = 2.52$) in water ($\varepsilon_{\mathrm{m}} = 1.77$), illuminated by a Gaussian beam with $w_0 = 5$ μm, power $P = 100$ mW, and wavelength $\lambda = 514$ nm. The resulting Rayleigh force (dashed green line) is dominated by the gradient force, allowing the formation of a stable trap. (b) Longitudinal components for $a = 5$ nm and $\varepsilon_{\mathrm{p}} < \varepsilon_{\mathrm{m}}$: The figure shows the same force components for a particle with $\varepsilon_{\mathrm{p}} < \varepsilon_{\mathrm{m}}$. In this case, the gradient force is repulsive, adding to the scattering force and accelerating the particle away from the optical axis, preventing the formation of a stable trap. (c) Longitudinal components for $a = 10$ nm and $\varepsilon_{\mathrm{p}} > \varepsilon_{\mathrm{m}}$: The figure shows the longitudinal components for a polystyrene particle with a radius of 10 nm ($\varepsilon_{\mathrm{p}} = 2.52$) in water ($\varepsilon_{\mathrm{m}} = 1.77$), illuminated by a Gaussian beam with $w_0 = 5$ μm, power $P = 100$ mW, and wavelength $\lambda = 514$ nm. The scattering force dominates the z-component of the gradient force, preventing the formation of a stable longitudinal trap. $\mathbf{r}_d = (0, 0, z/kw_0^2)$.

define the resultant of the longitudinal forces as the Rayleigh force (dashed green line), given by $\mathbf{F}_{\mathrm{Rayleigh}} = \mathbf{F}_{\mathrm{G}z} + \mathbf{F}_{\mathrm{scat}}$.

In figure 2.13(a), we present the longitudinal components as a function of the normalized longitudinal displacement for a spherical polystyrene particle with a radius of 5 nm, $\varepsilon_{\mathrm{p}} = 2.52$ in water, illuminated by a Gaussian beam with $w_0 = 5$ μm, power $P = 100$ mW, and wavelength $\lambda = 514$ nm. The scattering force reaches its maximum at $z = 0$, while the z-component of the gradient force attains its maximum at $z = kw_0^2/2\sqrt{3}$.

Note that $\mathbf{F}_{\text{Rayleigh}}$ is dominated by $\mathbf{F}_{Gz}$, which allows the particle to be trapped in a region just above the beam waist plane ($z = 0$), where $\mathbf{F}_{\text{Rayleigh}} = 0$. For a particle of this size and a beam with these parameters, the effect of the scattering force is perturbative. In other words, the gradient force is dominant and acts as a restoring force that pulls the particle back toward the plane $z = 0$, creating a stable trap. The scattering force, on the other hand, is relatively small and insufficient to overcome the gradient force. Thus, the scattering force has a perturbative effect but does not destabilize the optical trap created by the gradient force. Since in the transverse direction only the gradient force exists, which is restoring in this configuration, the particle experiences stable trapping in all three dimensions, allowing it to be manipulated both transversely and longitudinally.

Figure 2.13(b) depicts the same situation as in the previous case, but for $\varepsilon_p < \varepsilon_m$. In this scenario, although the gradient force is still dominant over the scattering force, trapping does not occur because the gradient force acts as a repulsive force. Consequently, the gradient force adds to the scattering force, accelerating the particle away from the optical axis.

The major challenge in achieving stable traps for weakly focused beams arises from the fact that scattering forces increase with the particle radius as a^6, whereas the gradient force grows as a^3. This limits the range of particle sizes for which $\mathbf{F}_{Gz} > \mathbf{F}_{\text{scat}}$, and a stable optical trap can be formed.

Figure 2.13(c) illustrates this effect, showing that, for a particle with $a = 10$ nm, the scattering force dominates overwhelmingly the z component of the gradient force. In this scenario, longitudinal trapping cannot occur. This is evidenced by the fact that the gradient force is insufficient to counteract the scattering force along the optical axis. Because the scattering force is dominant, it pushes the particle in the direction of propagation of the beam, preventing the formation of a stable trap.

We can define a criterion (R) for achieving axial stability. The criterion we will use was initially established by Ashkin [15] and later corrected by Harada [4]. The criterion for axial stability is based on the requirement that the magnitude of the z-component of the gradient force must be greater than the magnitude of the scattering force at the position where the gradient force reaches its maximum. That is,

$$R = \frac{|F_{Gz}(0, 0, kw_0^2/2\sqrt{3})|}{F_{\text{scat}}(0, 0, kw_0^2/2\sqrt{3})} \geqslant 1. \tag{2.94}$$

By analyzing equations (2.87) and (2.93), we verify that R is a function of λ, w_0, and a for a zero-order Gaussian beam, as presented by [4, 15]

$$R = R_0 \frac{1}{\left| \dfrac{\varepsilon_p - \varepsilon_m}{\varepsilon_p + 2\varepsilon_m} \right|} \frac{\lambda^5}{n_m^5 a^3 w_0^2} \geqslant 1, \tag{2.95}$$

where $R_0 \approx 1.3 \times 10^{-4}$. Equation (2.95) establishes a relationship between the beam parameters (λ, w_0) and the particle parameters (a, ε_p) that influence trap stability.

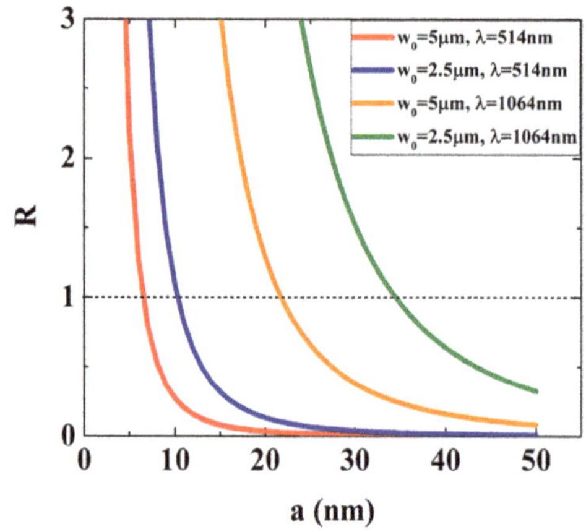

Figure 2.14. Axial stability criterion (R) as a function of particle radius. The figure shows the variation of the axial stability criterion R as a function of particle radius for different wavelengths ($\lambda = 514$ nm and $\lambda = 1064$ nm) and beam waists ($w_0 = 5\,\mu$m and $w_0 = 2.5\,\mu$m). Solid red line: $\lambda = 514$ nm and $w_0 = 5\,\mu$m. Solid blue line: $\lambda = 514$ nm and $w_0 = 2.5\,\mu$m. Solid orange line: $\lambda = 1064$ nm and $w_0 = 5\,\mu$m. Solid green line: $\lambda = 1064$ nm and $w_0 = 2.5\,\mu$m. Values of $R \geqslant 1$ indicate axial stability of the optical trap.

Applying this equation to the particle in figure 2.13, we obtain that, for $a = 5$ nm, $R = 2.17$, whereas for $a = 10$ nm, $R = 0.27$. This result is consistent with our graphical analysis, which indicates the existence of a stable trap only for $a = 5$ nm.

However, since $R \propto 1/w_0^2$, the $a = 10$ nm particle can be trapped if subjected to a more tightly focused beam, for example, with $w_0 = 2.5\ \mu$m, in which case $R \approx 1.1$. Another alternative for trapping the $a = 10$ nm particle is to exploit the dependence $R \propto \lambda^5$ and use a beam with a longer wavelength. Using $\lambda = 1064$ nm, we obtain $R = 10$.

Figure 2.14 illustrates the behavior of R as a function of the particle radius for two wavelengths ($\lambda = 514$ nm and $\lambda = 1064$ nm) and two beam waist values ($w_0 = 5\ \mu$m and $w_0 = 2.5\ \mu$m). The error associated with describing the fields as zero-order Gaussian beams is less than 3% when compared to the correction by Barton and Alexander [6], with the largest error corresponding to the beam with $\lambda = 1064$ nm and $w_0 = 2.5\ \mu$m, where $s = 0.025$, resulting in an error of approximately 2.4%. In this configuration, three-dimensional optical traps can be formed for particles with radii up to approximately 35 nm.

2.2.5 Validity range of the Rayleigh approximation

Our discussion so far has been based on the quasi-static approximation and the assumption that the fields are well described by the zero-order Gaussian approximation. However, how realistic are these approximations, and effectively, for what range of particle sizes are the optical forces well described by equations (2.85)–(2.87) and (2.93)? The goal of this section is to estimate the upper size limit for which the Rayleigh approximation is reasonably applicable.

Fortunately, this issue was addressed in a fundamental study published by Harada and Asakura [4], who investigated the range of particle sizes for which the Rayleigh approximation is valid in describing the radiation force. In this work, the authors graphically compare the longitudinal components ($\mathbf{F}_z$, Rayleigh approximation) and transversal components ($\mathbf{F}_x$, Rayleigh approximation) of the radiation force obtained using this approximation with those predicted by the generalized Lorenz–Mie theory (GLMT) [14], i.e. the longitudinal components ($\mathbf{F}_z$, GLMT) and transversal components ($\mathbf{F}_x$, GLMT).

In the Rayleigh approximation, the transversal component of the force is purely the gradient force, while the longitudinal component is the resultant of the scattering force and the gradient force. In the GLMT, both components (longitudinal and transversal) include both contributions. The results obtained by Harada and Asakura are shown in figure 2.15.

Figure 2.15(a) compares the results obtained by the Rayleigh approximation with those predicted by the GLMT theory for a polystyrene particle ($\varepsilon_d = 2.52$) in water ($\varepsilon_{\text{m}} = 1.77$), illuminated by a Gaussian beam with $\lambda = 514$ nm and waist $w_0 = 5\,\mu$m, considering a wide range of radii (1 nm to 10 μm). Note that the agreement between both models is very good for both components within the size range typically well described by Rayleigh scattering, that is, when $a < \lambda/20n_{\text{m}} \approx 20$ nm.

For larger radii, a growing discrepancy is observed in the longitudinal component of the force between the Rayleigh approximation and the GLMT, and this discrepancy increases as the particle radius grows. In contrast, the agreement of the transversal component between the models remains intact up to the so-called intermediate regime ($a \approx \lambda$), ensuring the validity of the Rayleigh approximation for obtaining transversal forces.

However, when the particle radius becomes comparable to or larger than the beam waist, the agreement in the transversal component between the models breaks

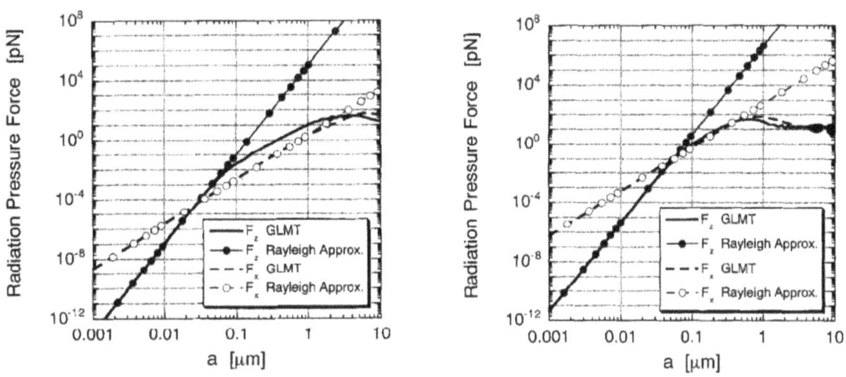

(a) Comparison between Rayleigh approximation and GLMT theory. $w_0 = 5\ \mu m$.

(b) Comparison between Rayleigh approximation and GLMT theory. $w_0 = 0.77\ \mu m$.

Figure 2.15. Comparison between longitudinal and transverse forces predicted by the Rayleigh and GLMT models as a function of particle radius. (a) Weakly focused beam $w_0 = 5\ \mu$m. (b) Strongly focused beam $w_0 = 0.77\,\mu$m. Solid black line: longitudinal component obtained by GLMT. Black dashed line: transverse component obtained by GLMT. Solid line with black dots: longitudinal component obtained using the Rayleigh approximation. Dashed line with white dots: transverse component obtained using the Rayleigh approximation. The longitudinal forces were calculated at position (0, 0, 0), while the transverse forces were calculated at position $(-w_0/2, 0, 0)$. Reprinted from [4], Copyright 1996, with permission from Elsevier.

down, with the Rayleigh approximation overestimating the force. This discrepancy occurs because, in GLMT, the a^3 dependence disappears in the transversal component when $a \geqslant w_0$. Furthermore, as discussed in section 2.2.3, the quasi-static approximation requires that $a < w_0$ for the field to be approximately uniform over the particle's volume in a non-uniform beam.

From figure 2.15(a), we can conclude that the upper limits for the validity of the Rayleigh approximation differ between the longitudinal and transversal components of the radiation force. The origin of this difference lies in the physical mechanisms that give rise to each of these forces. While the longitudinal component is dominated by the scattering force—resulting from the transfer of momentum from the EM wave—and is therefore subject to the limitations imposed by the type of scattering it undergoes (Rayleigh, Mie, etc)[10].

On the other hand, the transversal component is dominated by the gradient force, which originates from the electrostatic interaction of the incident field with the induced charges in the particle. As long as the description of the charge distribution is dipolar[11], i.e. the field is uniform over the volume of the particle, the transversal force can be described by the Rayleigh approximation (dipolar). This interval is larger because, for a Gaussian beam, the criterion for the uniformity of the electric field[12] is the parameter L, where $L \approx w_0 > \lambda/20n_m$ for a Gaussian beam. The discrepancy between the models for $a > w_0$ occurs because, in this regime, the particle can no longer be approximated as a dipole, and higher modes became dominant (quadrupole, octupole, etc), modifying the charge distribution and consequently the gradient force.

Note that what defines the upper limit is precisely the condition that satisfies the origin of each force component. That is, the particle must obey Rayleigh scattering (for the longitudinal component) and the induced charges must be described as a dipole (transversal component). This can be seen more clearly in the case of illumination by a highly focused Gaussian beam, as shown in figure 2.15(b), where $w_0 = 2\lambda/n_m = 770$ nm, where $s = 0.08 \ll 1$, and therefore still well described by the zero-order approximation. Observe that the upper particle size limit is shifted to a smaller value of $a \approx w_0$ in the transversal component, while the size limit for the longitudinal component remains at the same value as obtained in figure 2.15(a).

The result obtained by Harada and Asakura encourages us to use the dipolar gradient force approximation (Rayleigh) to obtain the transversal component of the

[10] Since the self-interaction term of the polarizability depends on the scattering, the description in terms of the Lorentz–Lorenz relation is more appropriate to obtain the transverse component as we can observe in Harada's work.

[11] What we are effectively stating here is that the incident field 'sees' an approximately dipolar surface charge distribution in the particle during the electrostatic interaction. This does not mean that higher-order excitation modes, such as the quadrupole, do not exist; however, even if they are present, they are not the dominant excitation modes when the field is approximately constant over the volume of the particle.

[12] The change in the consideration of the uniformity of the electric field from $a \ll \lambda$, formulated assuming interaction with a plane wave, to $a \leqslant w_0$ occurs because, in a Gaussian beam, the variation of the electric field is more gradual along the waist of the beam w_0 than along the wavelength. Therefore, for particles with $a \leqslant w_0$, the electric field can be treated as approximately uniform.

optical force within a particle size range where $a \leqslant w_0$, whenever the gradient force is the dominant force in that direction. The simplicity of the equations involved, together with the good agreement with the GLMT, justifies the use of this model as a first approximation for the study of various optical systems. In particular, later in this book we will use this model to study optical forces on plasmonic particles and semiconductor particles.

2.2.6 Spin–curl force

Finally, the last term in equation (2.71) represents the spin–curl force. This force arises due to polarization gradients in the EM field and is non-conservative. For this force to occur, the field polarization must be inhomogeneous[13]. The magnitude of $\mathbf{F}_{\text{spin}}$ is relatively small compared to the gradient and extinction forces and, therefore, it generally does not play a significant role in optical trapping experiments.

The expression for the spin-curl force is given by

$$\mathbf{F}_{\text{spin}}(\mathbf{r_d}) = -\frac{1}{2}\sigma_{\text{ext}}\nabla \times \mathbf{s}_d(\mathbf{r}_d), \tag{2.96}$$

where σ_{ext} is the extinction cross-section and $\mathbf{s}_d$ is the spin density, given by

$$\mathbf{s}_d = \frac{i\varepsilon_0\varepsilon_{\text{m}}}{2\omega}\mathbf{E}_i \times \mathbf{E}_i^*. \tag{2.97}$$

The spin density $\mathbf{s}_d$ in an EM field describes the spatial distribution of the intrinsic angular momentum associated with the field polarization. Simply put, it measures how the electric and magnetic fields 'rotate' around a point in space.

Broadly speaking, the spin–curl force is a term that quantifies the force exerted by the intrinsic angular momentum of light on a particle. The effect of this force on the particle depends on how $\mathbf{s}_d$ is generated. For circularly polarized light, this force induces an angular momentum in the particle, causing it to rotate around its own axis. In light fields with complex polarization gradients, such as radial or azimuthal polarization, the spin–curl force can drive the particle into an orbital motion around the optical axis of the light beam. The spin–curl force vanishes for a linearly polarized EM field since the spin density is zero in this case.

References

[1] Jackson J D 1998 *Classical electrodynamics.* (New York: Wiley)

[2] Jones P H, Maragò O M and Volpe G 2015 *Optical tweezers: Principles and applications.* (Cambridge: Cambridge University Press)

[3] van de Hulst H C 1981 *Light Scattering by Small Particles* (Courier Corporation)

[13] Non-uniform polarization refers to a situation where the polarization of the EM field varies from point to point in space. In other words, the direction and/or magnitude of the electric field polarization vector are not constant throughout space. Examples of inhomogeneous polarization include circular, radial, azimuthal, and elliptical polarization.

[4] Harada Y and Asakura T 1996 Radiation forces on a dielectric sphere in the rayleigh scattering regime *Optics Commun.* **124** 529
[5] Davis L W 1979 Theory of electromagnetic beams *Phys. Rev.* A **19** 1177
[6] Barton J P and Alexander D R 1989 Fifth-order corrected electromagnetic field components for a fundamental gaussian beam *J. Appl. Phys.* **66** 2800
[7] Barton J P, Alexander D R and Schaub S A 1988 Internal and near-surface electromagnetic fields for a spherical particle irradiated by a focused laser beam *J. Appl. Phys.* **64** 1632
[8] So J and Choi J-M 2016 Tuning the stiffness asymmetry of optical tweezers via polarization control *J. Korean Phys. Soc.* **68** 762
[9] Moura T A, Andrade U M S, Mendes J B S and Rocha M S 2020 Silicon microparticles as handles for optical tweezers experiments *Opt. Lett.* **45** 1055
[10] Oliveira K M, Moura T A, Lucas J L C, Teixeira A V N C, Rocha M S and Mendes J B S 2023 Use of organic semiconductors as handles for optical tweezers experiments: trapping and manipulating polyaniline (PANI) microparticles *ACS Appl. Polym. Mater.* **5** 3912
[11] Moura T A, Lana Júnior M L, da Silva C H V, Américo L R, Mendes J B S, Brandão M C N P, Subtil A G S and Rocha M S 2024 Optical trapping and manipulation of fluorescent polymer-based nanostructures: measuring optical properties of materials in the nanoscale range *Phys. Rev. Appl.* **22** 064043
[12] Campos W H, Moura T A, Marques O J B J, Fonseca J M, Moura-Melo W A, Rocha M S and Mendes J B S 2019 Germanium microparticles as optically induced oscillators in optical tweezers *Phys. Rev. Res.* **1** 033119
[13] Moura T A, Andrade U M S, Mendes J B S and Rocha M S 2023 Modulating the trapping and manipulation of semiconductor particles using bessel beam optical tweezers *Optics Lasers Eng* **170** 107778
[14] Gouesbet G, Maheu B and Gréhan G 1988 Light scattering from a sphere arbitrarily located in a gaussian beam, using a Bromwich formulation *J. Opt. Soc. Am.* A **5** 1427
[15] Ashkin A, Dziedzic J M, Bjorkholm J E and Chu S 1986 Observation of a single-beam gradient force optical trap for dielectric particles *Opt. Lett.* **11** 288

Chapter 3

The geometrical optics regime

In this chapter, we present and discuss a basic geometrical optics (GO) formalism to calculate the optical forces on spherical-shaped beads, valid when these beads have sizes considerably higher than the wavelength employed in the tweezers ($a \gg \lambda$). The goal is to give the reader insights on how to implement and, if necessary, generalize this type of calculation for different specific situations. We start from the force exerted by a single ray and then sum (integrate) for the entire laser beam, considering its geometry. Furthermore, laser absorption by the beads is also considered and included in the calculations, allowing one to analyze its effects in dielectrics with non-negligible absorption coefficients.

3.1 Optical forces in the geometrical optics regime

Many authors developed GO models in the past years and have their own approaches and algorithms to calculate the gradient and scattering forces on the beads in the GO regime [1–4], presenting equivalent results. Below we discuss the approach presented in reference [3], based on Ashkin's original work [2], which is didactically appropriate to the purposes of this book. In summary, we start calculating the force exerted by a single ray on a bead and then sum (integrate) over the entire laser beam, taking into account the highly focused geometry of the beam and its specific intensity profile.

3.1.1 Force exerted by a single ray

In figure 3.1, we represent a single ray of light with power dP hitting the surface of a bead with an incidence angle α measured with respect to the normal, as usual. We adopt an auxiliary coordinate frame in which the incidence direction of the ray is along the $\hat{z}'$ axis. When hitting the bead–medium interface three phenomena usually occur, with light being partially reflected, refracted and absorbed. Let us firstly neglect light absorption and consider only reflection and refraction, which is a good

doi:10.1088/978-0-7503-6074-6ch3

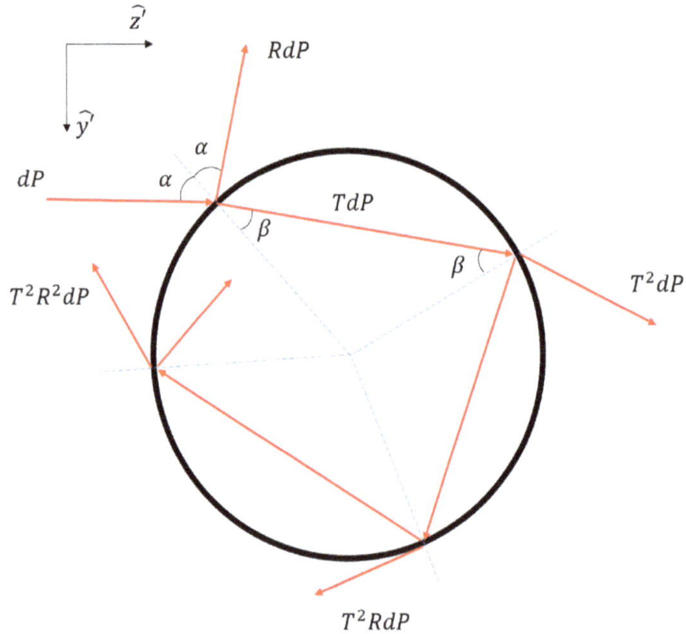

Figure 3.1. A single ray of light with power dP hitting the surface of a bead. At the bead–medium interface three phenomena usually occur, with light being partially reflected, refracted and absorbed. Let us first neglect light absorption and consider only reflection and refraction. Some reflected and refracted rays that the incident ray generates along its propagation are represented in the figure. Let R and T be, respectively, the reflectivity and transmissivity at the bead–medium interface. We label each specific ray by its corresponding power, which is determined by the number of reflections and refractions that it undergoes.

approximation for most semi-transparent dielectric materials. Some reflected and refracted rays that the incident ray generates along its propagation are represented in the figure. The angles of incidence (α) and refraction (β) are also represented. Let R and T be, respectively, the reflectivity and transmissivity at the bead–medium interface. We label each specific ray in figure 3.1 by its corresponding power, which is determined by the number of reflections and refractions that it undergoes, as illustrated.

The resulting optical force exerted by a single ray on the bead was first calculated by Roosen [1] and Ashkin [2]. These authors performed such calculations by adding the infinite series of multiple reflections and refractions represented in figure 3.1, which can be used to obtain the changes on the bead linear momentum. Thus, the optical force $d\mathbf{F}$ on the bead can be calculated by directly applying Newton's second law. The final result can be written as [1, 2]

$$d\mathbf{F} = \frac{n_{\rm m}}{c}[\mathrm{Re}(Q_{\rm t})\hat{z}' + Im(Q_{\rm t})\hat{y}']dP, \tag{3.1}$$

where $n_{\rm m}$ is the refractive index of the medium surrounding the bead, c is the velocity of light in vacuum, $\hat{z}'$ and $\hat{y}'$ are the unit vectors indicated in figure 3.1, and

$$Q_t = 1 + R\exp(2i\alpha) - T^2\frac{\exp[2i(\alpha-\beta)]}{1+R\exp(-2i\beta)}. \tag{3.2}$$

The expression for the effective reflectivity R can be appropriately written as an average over the two possible polarization states, TE and TM [5],

$$R(\alpha, \beta) = \frac{1}{2}\left[\frac{\sin(\alpha-\beta)}{\sin(\alpha+\beta)}\right]^2 + \frac{1}{2}\left[\frac{\tan(\alpha-\beta)}{\tan(\alpha+\beta)}\right]^2 \tag{3.3}$$

and for the transmissivity one can use

$$T(\alpha, \beta) = 1 - R(\alpha, \beta). \tag{3.4}$$

With this result for a single ray, the optical force exerted by a focused beam can be obtained by adding the contribution coming from each ray of the beam, i.e. by integrating equation (3.1) over the volume of the beam. Such a task will be discussed in the following section.

3.1.2 Force exerted by a focused beam

The total optical force exerted by a beam on the bead can be straightforwardly written from equation (3.1) as

$$\mathbf{F} = \int d\mathbf{F}. \tag{3.5}$$

In order to perform such integration, one must know some characteristics of the laser beam and of the objective used to focus it. Most setups use Gaussian (TEM_{00}) beam profiles, whose the intensity *before focusing* (I) can be conveniently represented by the function

$$I(\rho) = I_0 \exp\left(\frac{-2\rho^2}{w^2}\right), \tag{3.6}$$

where ρ is the radial distance from the center of the beam, I_0 is the intensity of the beam at its center ($\rho = 0$) and w is the beam waist, which usually depends on the beam position along its propagation direction due to diffractive effects. For applying the current model, one needs to measure the beam waist at the back aperture of the objective lens, i.e., immediately before being focused. Thus, here we consider w a constant measured at the objective back aperture.

The power element dP of equation (3.1) can be written in cylindrical coordinates as

$$dP = I(\rho)\rho d\rho d\varphi, \tag{3.7}$$

where φ is the azimuthal angle.

In figure 3.2 we represent the geometry of the beam after being focused by the objective lens. Observe that we adopt the origin of our coordinate frame xyz at the

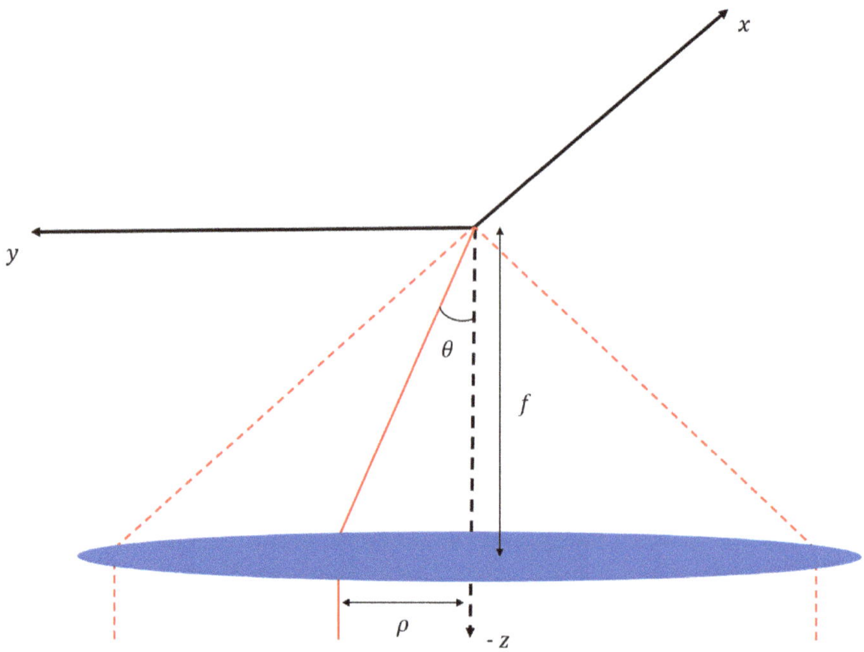

Figure 3.2. Geometry of the beam being focused by the objective lens. Observe that we adopt the origin of our coordinate frame xyz at the geometrical focus, such that the lens is localized in the xy-plane, at $z = -f$, the focal distance. We label a particular ray by its radial position ρ before being focused, and by the angle θ illustrated in the figure after it passes though the lens.

geometrical focus, such that the lens is localized in the xy-plane, at $z = -f$, the focal distance. We label a particular ray by its radial position ρ before being focused, and by the angle θ illustrated in the figure after it passes though the lens.

To properly describe the rays focusing in modern objectives, one should use the Abbe sine condition [6, 7], given by

$$\rho = f \sin\theta, \tag{3.8}$$

which is valid for a modern objective lens.

In addition, the focal distance of the objective can be conveniently expressed as

$$f = \frac{n_g R}{NA}, \tag{3.9}$$

where n_g is the refractive index of the glass used in the lens fabrication, R is the radius of the objective back aperture, and NA is its numerical aperture.

The total power of the beam before reaching the objective back aperture (P_t) can be expressed as

$$P_t = \int_0^{2\pi} \int_0^{\infty} I_0 \exp\left(\frac{-2\rho^2}{w^2}\right) \rho d\rho d\varphi. \tag{3.10}$$

Using equations (3.10), (3.8) and (3.7), one can finally write the power element dP as

$$dP = \frac{2P_t T_{\text{obj}}}{\pi w^2} \exp\left(\frac{-2f^2 \sin^2 \theta}{w^2}\right) f^2 \sin\theta \cos\theta d\theta d\varphi, \tag{3.11}$$

where T_{obj}, the *effective transmittance of the objective lens*, was purposely included in this expression to account for the power loss in the objective back aperture and inside the lens [8, 9].

Finally, equation (3.11) can be plugged into equation (3.1), and the total force exerted by the beam on the bead (equation (3.5)) can be rewritten as

$$\mathbf{F} = \int_0^{2\pi} \int_0^{\theta_0} d\mathbf{F}, \tag{3.12}$$

where θ_0 is the maximum value of θ.

This integral cannot be solved analytically for a closed result, but numerical calculations can be easily performed to determine the force knowing the parameters of interest. To proceed we need to express the angles of incidence (α), refraction (β) and the unitary vectors $\hat{z}'$, $\hat{y}'$ in equation (3.1) as a function of known variables to perform the integration. This task can be accomplished with the help of figure 3.3, in which we represent a bead trapped by a focused laser beam. Without loss of generality, let us assume that the equilibrium position occurs in the xz-plane, as illustrated in the figure. We label $\mathbf{r}_{\text{eq}}$ the vector that denotes the equilibrium position of the bead, i.e. the vector that goes from the origin of the coordinate frame (xyz) to the center of the trapped bead. This vector forms an angle γ with the z-axis as depicted. $\mathbf{a}$ is a vector whose magnitude is the bead radius a. Finally, $\mathbf{d}$ is a variable vector that goes from the origin to the surface of the bead, being anti-parallel to the incoming rays of light, as depicted in the figure for a particular (red colored) ray.

We start writing $\mathbf{d}$ in spherical coordinates,

$$\mathbf{d} = (d \sin\theta \cos\varphi, d \sin\theta \sin\varphi, d \cos\theta). \tag{3.13}$$

On the other hand, the equilibrium position of the bead ($\mathbf{r}_{\text{eq}}$) can be expressed as

$$\mathbf{r}_{\text{eq}} = (r_{\text{eq}} \sin\gamma, 0, r_{\text{eq}} \cos\gamma). \tag{3.14}$$

To write $\mathbf{d}$ as a function of r_{eq}, γ, θ and φ, we note that

$$|\mathbf{d} - \mathbf{r}_{\text{eq}}| = a, \tag{3.15}$$

which leads to

$$d^2 + r_{\text{eq}}^2 - 2r_{\text{eq}} d(\sin\gamma \sin\theta \cos\varphi + \cos\gamma \cos\theta) = a^2. \tag{3.16}$$

Solving this equation for d, we obtain

$$d = \sqrt{a^2 - r_{\text{eq}}^2 + r_{\text{eq}}^2(\sin\gamma \sin\theta \cos\varphi + \cos\gamma \cos\theta)^2} + r_{\text{eq}}(\sin\gamma \sin\theta \cos\varphi + \cos\gamma \cos\theta). \tag{3.17}$$

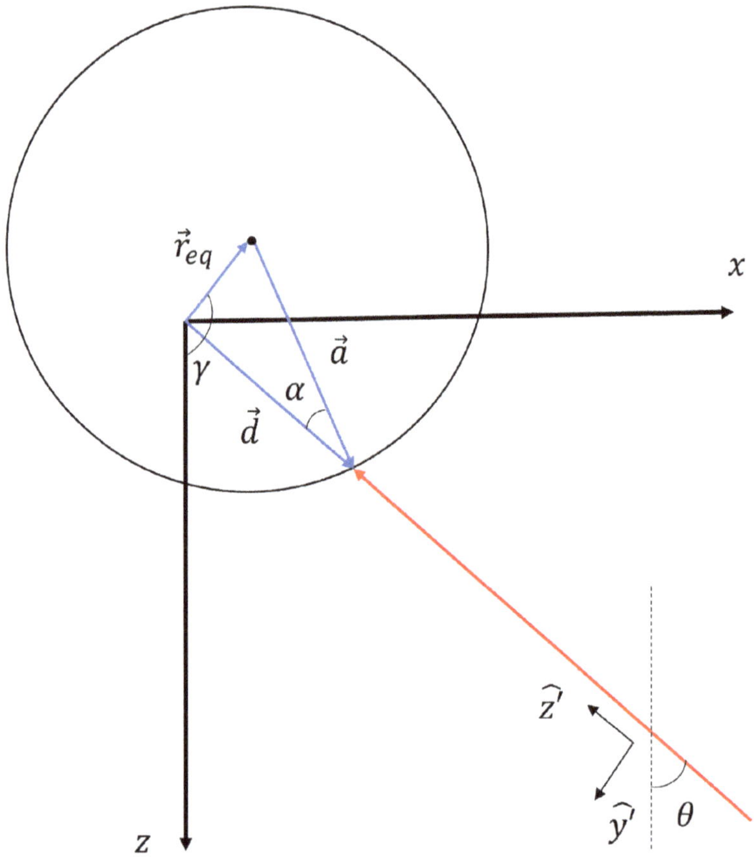

Figure 3.3. A bead trapped by a focused laser beam and the relevant variables used to perform the GO calculations. We label $\mathbf{r}_{eq}$ the vector that denotes the equilibrium position of the bead, i.e. the vector that goes from the origin of the coordinate frame (xyz) to the center of the trapped bead, forming an angle γ with the z-axis. $\mathbf{a}$ is a vector whose magnitude is the bead radius a. $\mathbf{d}$ is a variable vector that goes from the origin to the surface of the bead, being anti-parallel to the incoming rays.

Furthermore, using the triangle formed by the vectors $\mathbf{d}$, $\mathbf{r}_{eq}$ and $\mathbf{a}$, we get

$$r_{eq}^2 = d^2 + a^2 - 2ad\cos\alpha, \tag{3.18}$$

which leads to

$$\alpha = \arccos\left[\frac{d^2 + a^2 - r_{eq}^2}{2ad}\right] \tag{3.19}$$

The refraction angle β can be straightforwardly obtained using Snell's law,

$$n_m \sin\alpha = n_b \sin\beta, \tag{3.20}$$

where n_b is the bead refractive index.

Finally, to get the unit vectors $\hat{z}'$ and $\hat{y}'$, we first observe that $\hat{z}'$ is anti-parallel to **d**. Thus, using equation (3.13), we obtain

$$\hat{z}' = (-\sin\theta\cos\varphi, -\sin\theta\sin\varphi, -\cos\theta). \tag{3.21}$$

And by noting that the unit vector $\hat{y}'$ is parallel to the vector $\hat{z}' \times (\mathbf{r}_{eq} \times \hat{z}')$, we write.

$$\hat{y}' = \frac{\hat{z}' \times (\mathbf{r}_{eq} \times \hat{z}')}{|\hat{z}' \times (\mathbf{r}_{eq} \times \hat{z}')|}. \tag{3.22}$$

These relations can be used along with equation (3.12) to calculate the optical force exerted by the beam on the bead.

3.2 Comparison with experiments

In order to apply the formalism above to calculate the optical forces, and consequently the trap stiffness, one needs the following input parameters to use in the formulas:

- The refractive indices of the bead (n_b) and surrounding medium (n_m).
- The bead radius (a).
- The laser waist (w) and power at the objective back aperture (P_t).
- The objective transmittance (T_{obj}).

All these parameters are usually known for a given setup or can be measured with accuracy [9]. It is worth noting that, by using these parameters, the force calculation approach described here is independent of any adjustable parameter, such that any comparison performed between numerical results achieved with this model and experimental data is absolute [3].

To illustrate the accuracy of the GO approach so far discussed here, in figure 3.4 we show a direct comparison between the results achieved with these calculations and a set of experimental data obtained using oil beads of various sizes in water. The figure shows the transverse trap stiffness κ_t normalized by the laser local power in the focus (P_L) as a function of the bead radius a. These measurements were performed using a laser with 832 nm wavelength. Observe that the agreement is very good even for beads not much larger than such wavelength (radius $a > 1$ μm).

This result indicates that GO calculations can usually work well provided that the beads are larger than the laser wavelength, and therefore can be used to predict the behavior of the trap stiffness with accuracy for many typical optical tweezers (OT) setups.

Finally, note that the behavior of the trap stiffness κ as a function of the bead radius a in the GO regime is completely different from that determined in the Rayleigh regime. For the latter, we have shown that κ increases proportionally to a^3 (see chapters 1 and 2). The GO calculations, on the other hand, show that in this regime κ presents a hyperbolic decrease with the bead radius, i.e.

$$\kappa \propto \frac{1}{a}. \tag{3.23}$$

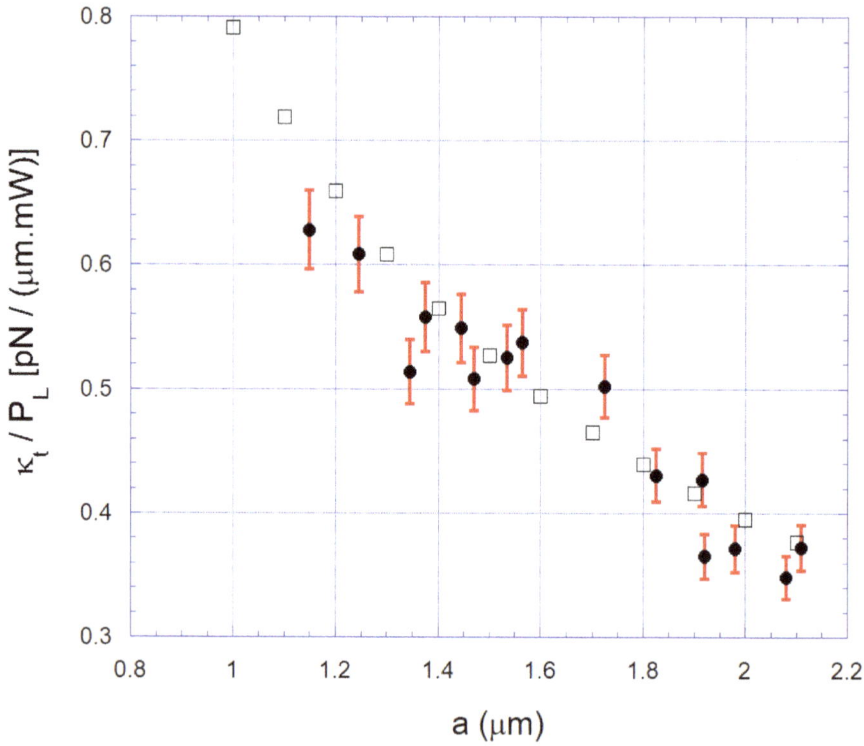

Figure 3.4. Transverse trap stiffness κ_t normalized by the laser local power in the focus (P_L) as a function of the bead radius a: a direct comparison between the results achieved with the GO approach presented here (*squares*) and a set of experimental data obtained using oil beads of various sizes in water (*circles*). These measurements were performed using a laser with 832 nm wavelength. Observe that the agreement is very good even for beads not much larger than such wavelength (radius $a > 1$ μm).

3.3 Including additional effects

3.3.1 Spherical aberration

The above calculations represent the core ideas behind this type of GO modeling. In fact, these calculations can be improved by adding effects such as spherical aberration, which degrades the focus and decreases the trapping efficiency [3]. Spherical aberration can be understood on the basis of figure 3.5: due to the refractive indices mismatch between the glass coverslip usually used to construct the sample chambers and the medium where the beads are immersed (usually water), the refracted rays do not converge to the same point, which means that the focal region will spread along the optical axis, decreasing the trap efficiency. Such an effect is stronger for smaller beads and also increases with the bead distance from the coverslip [3].

Reference [3] discusses in detail how to implement spherical aberration effects in the GO calculations presented here. Nevertheless, since in the GO regime the beads

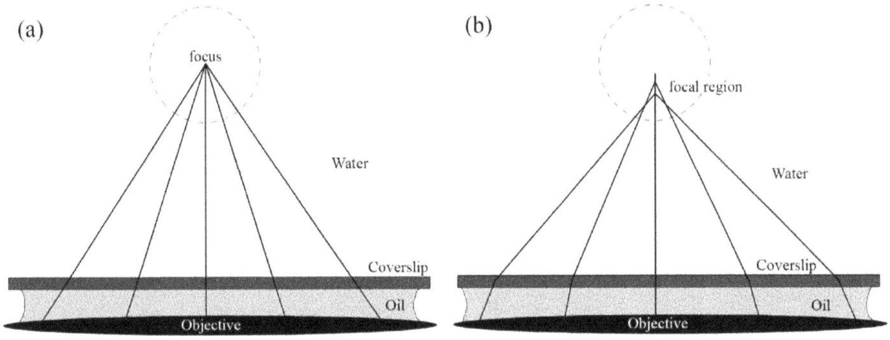

Figure 3.5. Spherical aberration in OT: due to the refractive indices mismatch between the glass coverslip usually used to construct the sample chambers and the medium where the beads are immersed (usually water), the refracted rays do not converge to the same point, which means that the focal region will spread along the optical axis, decreasing the trap efficiency. Note that this figure represents an oil immersion objective, much used in OT. The oil used should have the same refractive index of glass, avoiding additional losses related to light refraction. (a) Neglecting spherical aberration. (b) Considering spherical aberration. Reproduced [3], with permission from AIP Publishing.

are larger than the laser wavelength, the degradation of the trap efficiency is usually small for typical parameters used in most OT setups[1]. Figure 3.6 shows a comparison between GO calculations considering and not considering spherical aberration for oil beads in water, calculated using the same parameters of figure 3.5. Observe that the difference is really small, but increases for smaller bead radii. Furthermore, it is evident that spherical aberration decreases the values of the trap stiffness, i.e. degrades the trapping efficiency.

3.3.2 Light absorption by the bead

Another important effect, light absorption by the bead, can also be computed with accuracy in the GO framework; provided that such absorption is not too strong and can be described via the absorption coefficient of the material (ζ), as defined by the relation [10],

$$I = I_0 \exp(-\zeta l), \tag{3.24}$$

where I and I_0 are the transmitted and incident intensities measured at points separated by the distance l along the material.

In reference [10] the authors improved the GO formalism presented here to include light absorption during its propagation inside the bead. Basically, the same calculations can be done if equation (3.2) is generalized to [10]

$$Q_t = 1 + R\exp(2i\alpha) - T^2\frac{\exp[2i(\alpha - \beta)]}{\exp(2a\zeta\cos\beta) + R\exp(-2i\beta)}, \tag{3.25}$$

[1] For beads in the Rayleigh and in the intermediate regime, spherical aberration is usually much more critical and should be included in the calculations at least for oil immersion objectives [9].

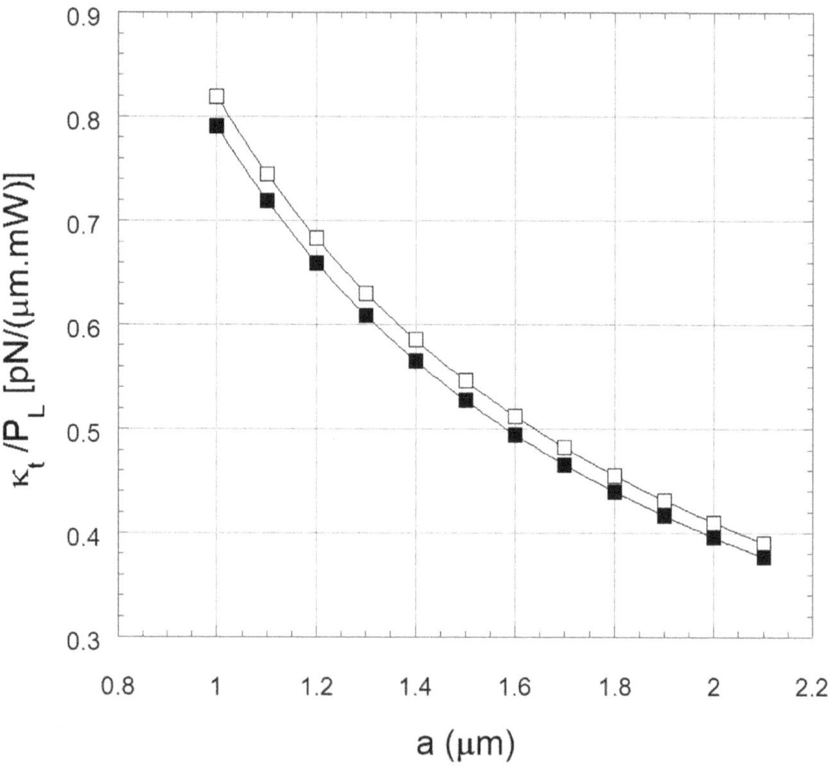

Figure 3.6. GO calculations of the transverse trap stiffness normalized by the laser local power in the focus (κ_t/P_L) considering (*black circles*) and not considering (*open circles*) spherical aberration for oil beads in water, calculated using the same parameters of figure 3.5. Observe that the difference is small, but increases for smaller bead radii. Furthermore, it is evident that spherical aberration decreases the values of the trap stiffness, i.e. degrades the trapping efficiency.

where ζ is the absorption coefficient of the bead defined above, a is the bead radius and all the other parameters are the same of equation (3.2). Note that if there is no absorption, $\zeta = 0$, and equation (3.25) will be equal to equation (3.2).

In reference [10], the authors compared the prediction of this model to experimental data obtained for bismuth telluride (Bi_2Te_3) beads in water. Such material presents a relatively small absorption coefficient at the laser wavelength used in the experiments (1064 nm) and thus the theory should work well. The comparison is shown in figure 3.7, in which the transverse radial component of the optical force is plotted as a function of the radial distance from the focus for a Bi_2Te_3 bead with radius $a = 3.1$ μm. Without absorption the theory fails to describe the experimental data for most values of ρ. When one considers absorption with a coefficient $\zeta = 0.11$ μm^{-1}, on the other hand, the agreement is much improved, showing that GO calculations can satisfactorily be used in this case to predict the behavior of the absorbing Bi_2Te_3 bead in an optical trap.

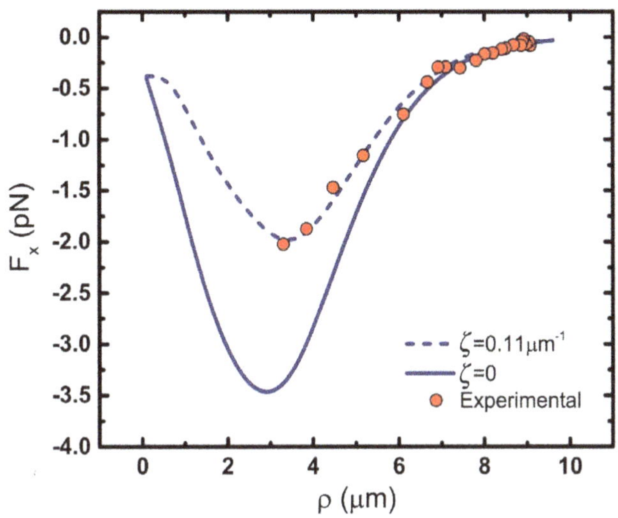

Figure 3.7. Radial component of the optical force as a function of the radial distance from the focus for a Bi_2Te_3 bead with radius $a = 3.1$ μm. Without absorption the theory fails to describe the experimental data for most values of ρ. When one considers absorption with a coefficient $\zeta = 0.11$ μm^{-1}, on the other hand, the agreement is much improved, showing that GO calculations can satisfactorily be used in this case to predict the behavior of the absorbing Bi_2Te_3 beads in an optical trap. Reprinted with permission from [10], copyright 2018 The Optical Society of America.

3.4 Non-Gaussian beams

The GO formalism presented here can also be adapted to other types of optical beams, for example, Bessel beams [11] or annular-shaped beams [12], which are important for studying other types of materials beyond semi-transparent dielectric beads, as will be evident in the following chapters.

Bessel beams, in particular, can present a series of advantages compared to Gaussian beams, allowing one to diminish the absorption of light from the focused beam by the beads due to its geometry [11]. In reference [11] in fact, the authors were capable to stably trap superparamagnetic beads composed by iron oxide dots in a polystyrene matrix using Bessel beams. Such trapping could not be achieved under similar conditions using a Gaussian beam [11].

In figure 3.8, we show a typical image of the Bessel beam used in reference [11] (panel (a)), its measured intensity profile along the diameter (panel (b)) and the typical behavior of the transverse trap stiffness as a function of the laser power for a 2.8 μm diameter superparamagnetic bead (panel (c)).

The measured intensity profile can be fitted to the function

$$I = I_0 J_0^2\left(\frac{2.4\rho}{\rho_B}\right), \tag{3.26}$$

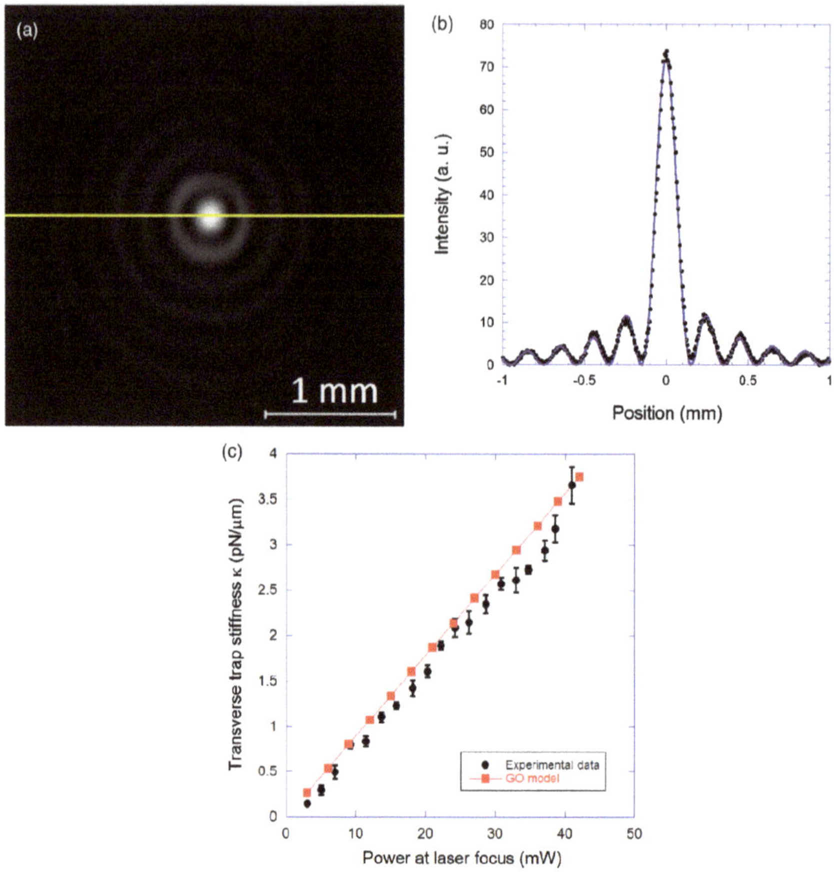

Figure 3.8. A typical image of the Bessel beam used in reference [11] (a), its measured intensity profile along the diameter (b) and the typical behavior of the transverse trap stiffness as a function of the laser power for a 2.8 μm diameter superparamagnetic bead (c). Reprinted with permission from [11], copyright 2018 The Optical Society of America.

where I is the intensity at a radial distance ρ from the beam center, I_0 is the intensity of the central peak (at $\rho = 0$), J_0 is the Bessel function of zeroth-order and ρ_B is the Bessel radius of the beam (distance from the center to the first minimum), an important characterization parameter. Such fitting is shown as a solid line in the panel (b) of figure 3.8 and allows one to quantitatively characterize the Bessel beam profile.

Finally, panel (c) of figure 3.8 shows the comparison between the prediction of the OG theory (*red squares*) and experimental data (*black circles*). The OG calculations were performed considering the non-negligible light absorption by the bead as discussed in the previous section, but in this case the absorption coefficient was measured by an independent technique, allowing an absolute comparison between theory and experiments, with excellent agreement.

To conclude the chapter, we emphasize that OG calculations are helpful in many experimental situations (different materials, different beam profiles), leading to an accurate description of the optical forces and trap stiffness provided that the beads are larger than the laser wavelength used in the OTs setup. The formalism is versatile and other effects not discussed here (e.g. other types of optical aberrations) can even be included, allowing a good description of the bead–laser interaction.

References

[1] Roosen G 1979 La lévitation optique de sphères *Can. J. Phys.* **57** 1260

[2] Ashkin A 1992 Forces of a single-beam gradient laser trap on a dielectric sphere in the ray optics regime *Biophys. J.* **61** 569

[3] Rocha M S 2009 Optical tweezers for undergraduates: theoretical analysis and experiments *Am. J. Phys.* **77** 704

[4] Callegari A, Mijalko A B G M and Volpe G 2015 Computational toolbox for optical tweezers in geometrical optics *J. Opt. Soc. Am.* B **32** B11

[5] Fowles G R 1975 *Introduction to Modern Optics* 2nd edn (New York: Dover)

[6] Maia Neto P A and Nussenzveig H M 2000 Theory of optical tweezers *Europhys. Lett.* **50** 702

[7] Mazolli A, Maia Neto P A and Nussenzveig H M 2003 Theory of trapping forces in optical tweezers *Proc. R. Soc. Lon.* A **459** 3021

[8] Viana N B, Rocha M S, Mesquita O N, Mazolli A and Maia Neto P A 2006 Characterization of objective transmittance for optical tweezers *Appl. Opt.* **45** 4263

[9] Viana N B, Rocha M S, Mesquita O N, Mazolli A, Maia Neto P A and Nussenzveig H M 2007 Towards absolute calibration of optical tweezers *Phys. Rev.* E **75** 021914

[10] Campos W H, Fonseca J M, Mendes J B S, Rocha M S and Moura-Melo W A 2018 How light absorption modifies the radiative force on a microparticle in optical tweezers *App. Opt.* **57** 7216

[11] Andrade U M S, Garcia A M and Rocha M S 2021 Bessel beam optical tweezers for manipulating superparamagnetic beads *Appl. Opt.* **60** 3422

[12] Oliveira L, Campos W H and Rocha M S 2018 Optical trapping and manipulation of superparamagnetic beads using annular-shaped beams *Methods Protoc* **1** 44

IOP Publishing

Optical Trapping and Manipulation of New Materials

Tiago de Assis Moura, Joaquim Bonfim Santos Mendes and Márcio Santos Rocha

Chapter 4

Metallic particles: plasmonics

Plasmonics can be defined as the part of photonics that studies the interaction between light and the collective oscillations of free electrons in conductive materials, such as metals. It primarily focuses on phenomena such as surface plasmons, which allow light to be confined and manipulated on subwavelength scales, with applications in sensors, integrated photonics, nanolithography, and optical forces. In this chapter, we will present a brief overview of plasmonics, with an emphasis on surface plasmon-polaritons (SPPs) and localized surface plasmons (LSPs). Then, we will discuss how optical forces manifest in the field of plasmonics, concluding with the introduction of the concept of plasmonic tweezers, which exploits SPPs to enhance the efficiency of optical trapping setups.

4.1 A brief introduction to plasmonics

A metal is characterized, among other aspects, by its high density of free electrons, which makes it a highly conductive medium. As a result, electromagnetic waves in the visible light range and lower frequencies do not propagate inside the metal due to the negative dielectric response that leads to strong reflection of the incident waves. Consequently, most of the intensity of the incident electromagnetic waves is reflected, causing the metal surface to act as a mirror. The fraction of the intensity that penetrates the metal is quickly attenuated over a short distance, known as the skin depth or penetration depth, which characterizes the depth at which the electromagnetic field decays due to energy dissipation in the metal.

However, metals can support electromagnetic waves in the visible range (and lower frequencies), provided that these waves travel confined to the surface of the metal (at the boundary with a dielectric medium) in dimensions smaller than the incident wavelength. In other words, the waves can propagate on, but not inside, conductive media as guided surface waves. These guided surface waves are known as SPPs and result from the coupling between electromagnetic fields and the collective oscillations of free electrons at the metal surface. These oscillations are longitudinal

doi:10.1088/978-0-7503-6074-6ch4

waves that oscillate at the same frequency as the surface wave propagates, being confined to the interface between the metal and the adjacent dielectric. The electromagnetic field associated with these waves decays exponentially both in the metal and in the adjacent dielectric, ensuring that the energy remains confined near the surface. This confinement leads to a significant increase in the local field, which enables the use of SPPs in numerous applications such as sensors, nanophotonics, and high-sensitivity optoelectronic devices.

In addition to the propagating SPPs, another form of plasma excitation can be induced in metallic nanostructures, known as LSPs. As the name suggests, these excitations are localized (non-propagating) at the surface of objects with dimensions much smaller than the wavelength of the incident radiation. The incident radiation couples to the free electrons of the nanostructure, while the curved surface of the nanostructure exerts an effective restoring force on the excited electrons, enabling a resonance. This leads to an amplification of the field both inside and in the near-field region outside the particle. This resonance is called localized surface plasmon resonance, or simply localized plasmon resonance. Another consequence of the curved surface is that plasmon resonances can be excited by direct light illumination, unlike propagating SPPs, which (as we will see) require phase-matching techniques to be excited.

4.1.1 Wave equation in conducting media

All the plasmonic effects we will discuss below originate from the way conductive media interact with electromagnetic fields. The starting point is the presence of free electric charges in conductive media, which generate an electric current density $\mathscr{J}$. Thus, when writing Maxwell's equations in a material medium without external sources, we must include $\mathscr{J}$ along with the displacement current (time derivative of the electric displacement $\mathscr{D}$) in the Ampère–Maxwell law. For a source-free conductive medium, Maxwell's equations take the form

$$\nabla \times \mathscr{H} = \frac{\partial \mathscr{D}}{\partial t} + \mathscr{J}, \tag{4.1}$$

$$\nabla \times \mathscr{E} = -\frac{\partial \mathscr{B}}{\partial t}, \tag{4.2}$$

$$\nabla \cdot \mathscr{B} = 0, \tag{4.3}$$

$$\nabla \cdot \mathscr{D} = 0. \tag{4.4}$$

For a monochromatic wave of frequency ω, the fields $\mathscr{H}$, $\mathscr{E}$, $\mathscr{B}$, and $\mathscr{D}$ can be written in complex form, for example, $\mathscr{H}(\mathbf{r}, t) = \mathbf{H}(\mathbf{r})e^{-i\omega t}$ (similarly for the other fields). Thus, we can rewrite equation (4.1) in the phasor form

$$\nabla \times \mathbf{H} = -i\omega \mathbf{D} + \mathbf{J}. \tag{4.5}$$

The conductive medium may exhibit dielectric properties (which can be attributed, for example, to bound electrons). In this case, the dielectric properties of the medium can be described in terms of its dielectric constant ε, such that

$$\mathbf{D} = \varepsilon\varepsilon_0\mathbf{E} = \varepsilon_0(1 + \chi)\mathbf{E}. \tag{4.6}$$

Similarly, the conductive properties of the medium can be described by the conductivity σ, allowing the current density to be expressed as a function of $\mathbf{E}$ by Ohm's law,

$$\mathbf{J} = \sigma\mathbf{E}. \tag{4.7}$$

Using equations (4.6) and (4.7) in equation (4.5), we have

$$\nabla \times \mathbf{H} = (-i\omega\varepsilon_0\varepsilon + \sigma)\mathbf{E} = -i\omega\varepsilon_0\left(\varepsilon + \frac{\sigma}{i\omega\varepsilon_0}\right)\mathbf{E},$$

or, in a compact form,

$$\nabla \times \mathbf{H} = -i\omega\varepsilon_0\varepsilon_c\mathbf{E}, \tag{4.8}$$

where

$$\varepsilon_c = \varepsilon + \frac{\sigma}{i\omega\varepsilon_0} \tag{4.9}$$

is the effective dielectric constant of the conductive medium. Note that ε_c is a complex, frequency-dependent parameter that combines the dielectric and conductive contributions of the medium. Thus, we have $\mathbf{D} = \varepsilon_0\varepsilon_c\mathbf{E}$ and $\mathbf{B} = \mu_0\mathbf{H}$, focusing our attention on non-magnetic conductive media.

The convenience of working with ε_c is that the wave equations for $\mathbf{E}$ and $\mathbf{H}$ retain the same structure as those obtained for dielectric media, so the solutions remain the same:

$$\nabla \times \nabla \times \mathbf{E} = -\nabla \times \frac{\partial \mathbf{B}}{\partial t},$$

$$\nabla \times \nabla \times \mathbf{E} = i\omega\mu_0\nabla \times \mathbf{H},$$

$$\nabla \times \nabla \times \mathbf{E} = \omega^2\mu_0\varepsilon_0\varepsilon_c\mathbf{E}.$$

Using $\nabla \times \nabla \times \mathbf{E} = \nabla(\nabla \cdot \mathbf{E}) - \nabla^2\mathbf{E}$, we have

$$\nabla(\nabla \cdot \mathbf{E}) - \nabla^2\mathbf{E} = \omega^2\mu_0\varepsilon_0\varepsilon_c\mathbf{E}. \tag{4.10}$$

Representing the operator ∇ as $i\mathbf{k}$, where $\mathbf{k}$ is the wave vector, equation (4.10) becomes

$$-\mathbf{k}(\mathbf{k} \cdot \mathbf{E}) + (\mathbf{k} \cdot \mathbf{k})\mathbf{E} = \frac{\omega^2}{c^2}\varepsilon_c\mathbf{E}. \tag{4.11}$$

There are two possible classes of solutions for equation (4.11): longitudinal waves, where **k** and **E** are parallel, and transverse waves, where **k** and **E** are orthogonal. Although longitudinal solutions are not possible in conventional dielectric media, in conductive media they can occur in the form of volume plasmons.

Volume plasmons are collective oscillations of free electrons within conductive materials. These longitudinal oscillations propagate through the medium, creating charge density fluctuations and associated electric fields. Characterized by a specific frequency, called the plasma frequency (ω_p), volume plasmons play a crucial role in the optical and electronic properties of conductive materials. However, volume plasmons cannot be directly excited by conventional light due to their longitudinal nature, contrasting with the transverse nature of electromagnetic waves, and the momentum mismatch between photons and plasmons. Alternative methods, such as electron pumping, can excite volume plasmons [1]. As this is a longitudinal solution, this topic will not be explored in depth in this book, but we recommend the references [1–3] for interested readers.

For transverse waves, we have $\mathbf{k} \cdot \mathbf{E} = 0$, and equation (4.11) reduces to

$$k^2\mathbf{E} = \frac{\omega^2}{c^2}\varepsilon_c(\omega)\mathbf{E}, \tag{4.12}$$

leading to the dispersion relation

$$k^2 = \frac{\omega^2}{c^2}\varepsilon_c(\omega). \tag{4.13}$$

All monochromatic propagating waves must obey equation (4.13). This dispersion relation is general, valid for any propagation medium, just replacing ε_c with the dielectric constant corresponding to the medium of interest. For example, in vacuum, where $\varepsilon_c = 1$, the relation becomes linear ($k = \omega/c$), defining the so-called 'light line' in an ω–k diagram.

For conducting media, using 4.9, we have

$$k^2 = \frac{\omega^2}{c^2}\varepsilon\left(1 + \frac{\sigma}{i\omega\varepsilon_0\varepsilon}\right). \tag{4.14}$$

Neglecting the dielectric contribution of the medium ($\varepsilon = 1$) and writing the conductivity in terms of Drude's conductivity $\sigma = \frac{\sigma_0}{1 + i\omega\tau}$, we have

$$\varepsilon_c = 1 - \frac{\sigma_0/\tau\varepsilon_0}{\omega^2 + i\omega/\tau},$$

where σ_0 is the low-frequency conductivity and τ is the scattering time (or collision time between free electrons). Writing $\omega_p = \sqrt{\sigma_0/\tau\varepsilon_0}$ and $\xi = 1/\tau$, we have

$$\varepsilon_c = 1 - \frac{\omega_p^2}{\omega^2 + i\omega\xi}, \tag{4.15}$$

where ξ is the collision rate and ω_p is the plasma frequency of the conducting medium[1]. In many situations of interest, we have $\omega \gg \xi$, such as in the case of metals in the visible and near IR spectrum, so we can discard the damping term and write (4.15) as

$$\varepsilon_c = 1 - \frac{\omega_p^2}{\omega^2}, \tag{4.16}$$

which is known as the simplified Drude model. It is clear from (4.16) that the existence of conductivity in the medium acts to suppress the electric permittivity to values lower than ε_0 when $\varepsilon_c < 1$. Moreover, using (4.16) in (4.14), we have

$$k^2 = \frac{\omega^2}{c^2}\left(1 - \frac{\omega_p^2}{\omega^2}\right),$$

$$\omega^2 = k^2c^2 + \omega_p^2. \tag{4.17}$$

Note that if $\omega < \omega_p$, the only possible solutions are those where k is purely imaginary, which corresponds to evanescent field solutions. Only for $\omega > \omega_p$ we have solutions in the form of propagating waves. In other words, when the frequency of the incident wave is lower than the plasma frequency of the conducting medium, the wavenumber inside the conducting medium is imaginary, corresponding to attenuation without propagation, so this spectral region can be treated as a forbidden band. Figure 4.1(a) illustrates the plasmonic dispersion in relation to the dispersion of the light line in a ω–k diagram. In the forbidden band, the dielectric constant takes negative values and the metal behaves as an SNG medium[2]. Figure 4.1(b) illustrates the behavior of the real part (dashed blue line) and imaginary part (solid green line) of ε_c given by equation (4.15) in comparison with the behavior of ε_c provided by the simplified Drude model (dashed red line) given by equation (4.16), as a function of frequency.

The plasma frequency of most metals is in the UV region of the spectrum. For example, the plasma frequency of gold ($\lambda_p = 139$ nm [4]), silver ($\lambda_p = 138$ nm [4]), copper ($\lambda_p = 142$ nm [4]), and aluminum ($\lambda_p = 103$ nm [4]), where $\lambda_p = \frac{2\pi c}{\omega_p}$ is the wavelength associated with the plasma frequency. Note that, for the visible and near IR regions of the spectrum, metals behave as SNG media with negative dielectric constant and, therefore, do not allow the propagation of waves with these frequencies. For waves with frequencies in the spectrum region above the UV

[1] Later we will present the formal definition of the plasma frequency for the reader who is not familiar with the concept.

[2] An SNG (single-negative medium) is characterized as a medium where one of the two parameters that describe the medium (electric permittivity and magnetic permeability) is negative. Since we are considering non-magnetic media with $\mu = 1$, SNG media are those whose dielectric constant $\varepsilon < 0$, so the electric permeability ($\varepsilon_0\varepsilon$) is also negative. Media where both parameters are negative are called DNG (double-negative medium) and are characterized as metamaterials. Media where both parameters are positive, called DPS (double-positive medium), behave like conventional dielectrics. More details can be found in reference [4].

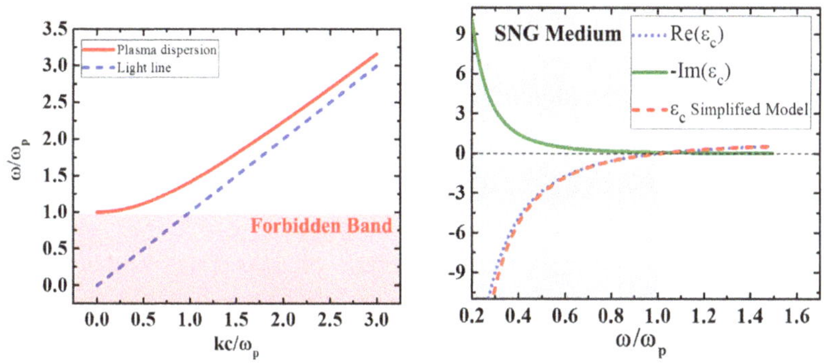

(a) The dispersion relation. (b) Effective dielectric constant in the Drude model

Figure 4.1. (a) Plasma dispersion relation (solid red line) and light line (dashed blue line). The red region indicates the forbidden band in metals. (b) Effective dielectric constant for metals described by the simplified Drude model (dashed red line) and the full Drude model (dashed blue line for the real part, solid green line for the imaginary part). The gray region indicates the frequency range where the metal behaves as an SNG medium.

(with the requirement that $\omega > \omega_p$), these media behave as ordinary dielectric media, allowing the existence of transverse propagating waves [1].

To better understand what the dispersion relation given by equation (4.17) means, let us analyze the parameters related to the solutions of propagating waves, i.e. the refractive index n, the wavenumber k, and the impedance of the medium η. These parameters can be directly derived from Maxwell's equations. Considering, for simplicity, that the solution of the wave equation is a plane wave, the complex amplitudes are written as

$$\mathbf{E}(\mathbf{r}) = \mathbf{E}_0 e^{i\mathbf{k}\cdot\mathbf{r}} \tag{4.18}$$

and

$$\mathbf{H}(\mathbf{r}) = \mathbf{H}_0 e^{i\mathbf{k}\cdot\mathbf{r}}, \tag{4.19}$$

where $\mathbf{E}_0$ and $\mathbf{H}_0$ are constant vectors, so that all components of $\mathbf{E}(\mathbf{r})$ and $\mathbf{H}(\mathbf{r})$ satisfy the Helmholtz equation. To satisfy Maxwell's equations, $\mathbf{E}_0$ and $\mathbf{H}_0$ must obey the relations

$$\mathbf{k} \times \mathbf{H}_0 = -\omega\varepsilon_0\varepsilon_c \mathbf{E}_0, \tag{4.20}$$

$$\mathbf{k} \times \mathbf{E}_0 = \omega\mu_0\mu \mathbf{H}_0. \tag{4.21}$$

Since the solutions are transverse waves, i.e. the vectors $\mathbf{E}_0$, $\mathbf{H}_0$, and $\mathbf{k}$ are mutually orthogonal and $H_0 = \frac{k}{\omega\mu_0\mu}E_0$, we can obtain from equation (4.20) or (4.21) the following relation:

$$k = \frac{\omega}{c}\sqrt{\mu\varepsilon_c} = k_0\sqrt{\varepsilon_c}, \tag{4.22}$$

where $k_0 = \frac{\omega}{c}$ is the wavenumber in vacuum, and we are considering non-magnetic media, so that $\mu = 1$. The impedance of the medium is written as the ratio between the electric field and magnetic field amplitudes, i.e.

$$\eta = \frac{E_0}{H_0} = \sqrt{\frac{\mu_0}{\varepsilon_0 \varepsilon_c}} = \frac{\eta_0}{\sqrt{\varepsilon_c}} = \frac{\eta_0}{n}, \tag{4.23}$$

where $\eta_0 = \sqrt{\mu_0/\varepsilon_0} = 377\ \Omega$ is the impedance of free space and $n = \sqrt{\varepsilon_c}$ is the refractive index of the medium. The intensity of the wave in the medium can be written in terms of the impedance of the medium as

$$I = \mathrm{Re}\left[\frac{1}{2}E_0 H_0^*\right] = \mathrm{Re}\left[\frac{|E_0|^2}{2\eta}\right]. \tag{4.24}$$

Considering that ε_c is given by equation (4.16), for $\omega < \omega_p$, these parameters become

$$k = i\frac{\omega}{c}\sqrt{|\varepsilon_c|}; \quad n = i\sqrt{|\varepsilon_c|}; \quad \eta = -i\frac{\eta_0}{\sqrt{|\varepsilon_c|}}, \quad I = 0. \tag{4.25}$$

Note that, for an ideal conducting medium with ε_c given by equation (4.16), the wavenumber, refractive index, and impedance are purely imaginary numbers, which leads to the fact that no power can be transported through a conducting medium ($I = 0$). But, if no power can be transported by the conducting medium, what happens to the power carried by an incident wave coming from an ordinary dielectric medium (DPS medium) when it strikes a conducting medium (SNG medium)?

Since the wave from the dielectric medium (medium 1) cannot propagate in the conducting medium (medium 2), when it strikes the boundary separating the two media, the electromagnetic wave is reflected. This becomes clearer when looking at the impedances of the two media, where the impedance of the dielectric medium (η_1) is real and the impedance of the conducting medium (η_2) is imaginary, which leads us to a complex reflectivity $|R| = 1$. The argument of R gives us the phase shift between the incident and reflected waves, while the magnitude of R gives us the fraction of the incident wave's power that is reflected. The reflection coefficient R in terms of the impedance of the media is given by the equation

$$R = \frac{\eta_2 - \eta_1}{\eta_2 + \eta_1}. \tag{4.26}$$

For real conductors, this reflection is accompanied by an evanescent field that decays exponentially within the conducting medium (as illustrated in figure 4.2), which can be easily demonstrated by applying the boundary conditions at the dielectric/conductor interface. This evanescent field is characterized by a parameter called the penetration depth (or skin depth) d_p, which gives us the distance from the dielectric/conductor interface at which the electromagnetic fields are attenuated to

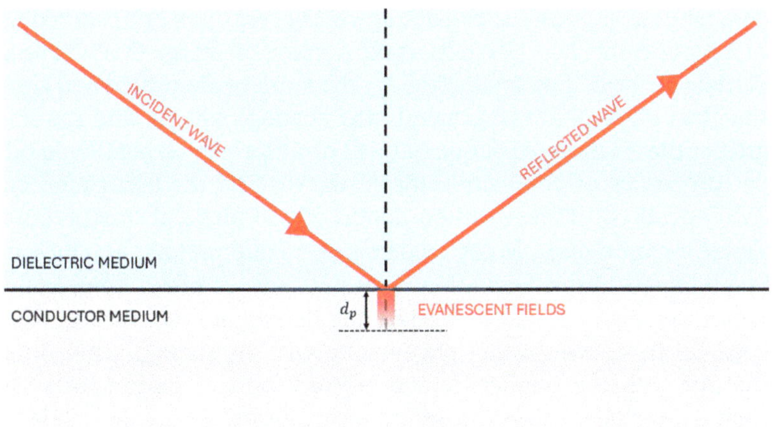

Figure 4.2. Scheme illustrating the emergence of evanescent fields and the reflection of an incident electromagnetic wave with $\omega < \omega_p$ at the interface between a dielectric medium and a conducting medium.

e^{-1} of their initial value. By writing the wavenumber as $k = k_1 + ik_2$, we can express $d_P = 1/k_2$, which, according to equation (4.25), leads to

$$d_p = \frac{c}{\omega\sqrt{|\varepsilon_c|}}. \tag{4.27}$$

Although the evanescent field decays exponentially within the conductor, it still carries energy locally, which is gradually dissipated as heat due to the absorption associated with the resistive properties of the material. Note that we used the term 'evanescent field', not 'evanescent wave', to describe the electromagnetic field within the material. This choice emphasizes that there is no wave propagating (in the classical sense) within the conducting medium, as occurs in dielectric media or in vacuum.

To make this clearer, we can generalize equation (4.17) by considering a complex wavenumber, which characterizes the presence of an evanescent wave. However, to satisfy the dispersion relation in a conductor, it would be necessary for $k_2 \gg k_1$ when $\omega < \omega_p$. This would result in such strong attenuation that $d_p \ll \lambda$, i.e. the field would be so strongly damped that it would not complete even one oscillation inside the material. Thus, the energy of the incident wave does not propagate within the conductor. It is stored locally in the regions where the field exists (i.e. within the penetration depth d_p) and is rapidly absorbed and converted into heat due to resistive processes. Thus, our definition of 'forbidden band', which we initially defined for ideal conductors, remains valid for real conductors, since no electromagnetic wave actually propagates within a conductor, whether it is real or ideal.

As an exercise for the reader, derive the parameters of equation (4.25), writing ε_c from equation (4.15) (or from equation (4.9)), where we include the dissipation term $i\omega\xi$. In this scenario, the parameters k, n, and η take on complex values (with a real part), so that we will have $I \neq 0$ inside the conductor, which, as mentioned earlier, is related to the evanescent field. Another change that the reader will notice is that, once η is complex, we will have $|R| < 1$, precisely to account for the absorbed power.

Note that while d_p tells us the distance at which energy will be absorbed within the conductor, $|R|$ (or rather $1 - |R|$) tells us the fraction of energy that will be absorbed and converted into heat. The smaller the d_p, the more concentrated and closer to the surface the absorption (and consequently the heating) will be, and the smaller the $|R|$, the greater the fraction of energy that will be absorbed. Generally, in tables with the optical properties of conducting materials, we find the absorption coefficient $\alpha_{ab} = 1/d_p$ instead of the penetration depth, whose physical interpretation arises from the interpretation of d_p. Many readers tend to interpret α_{ab} as a measure of the material's ability to absorb radiation; however, this is not a completely correct interpretation, since, as we stated, it is $|R|$ that determines the fraction of energy to be absorbed. In fact, two distinct conductors with the same α_{ab} but different $|R|$ values, the one with the smaller $|R|$ will absorb more radiation. Moreover, it is possible that α_{ab} includes other absorption mechanisms besides those related to free electrons, such as, for example, through electronic transitions. In this case, not all of the absorbed energy will be converted into heat, and we will address this topic later when we discuss semiconductors, where these mechanisms play an important role.

4.2 Surface plasma polaritons (SPP)

Under certain circumstances, as we will see in this section, a surface wave can be generated at the interface between a dielectric medium and a conductor. This surface wave, known as an SPP, propagates along the interface, exhibiting an evanescent nature in both adjacent media.

Let us assume that medium 1 is a dielectric with dielectric constant ε_1, while medium 2 is a conductor with dielectric constant ε_2. Both media are assumed to be non-magnetic ($\mu_1 = \mu_2 = 1$). To simplify the analysis, we further assume that the media are ideal and lossless, so that ε_1 and ε_2 are real numbers. Another simplification is to consider the media as semi-infinite planes, with the dielectric medium extending over the xz plane for $y > 0$, and the conductor extending over the zx plane for $y < 0$. Figure 4.3 illustrates the initial arrangement and the coordinate system used.

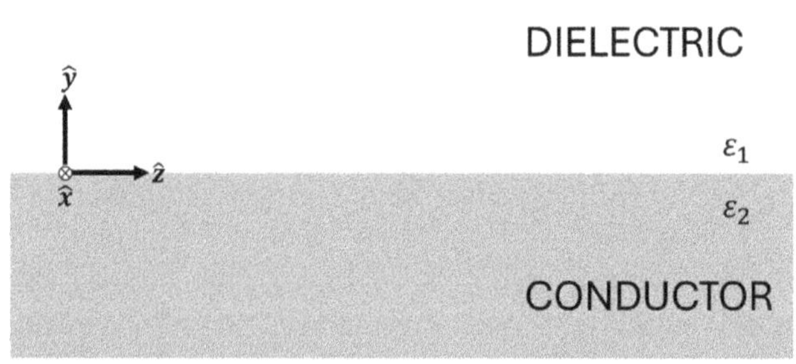

Figure 4.3. Illustration of the initial arrangement and coordinate system used for the analysis of SPPs.

We will begin by demonstrating that the interface between the media can support a propagating surface wave. The first condition we impose is that the surface wave should be a TM wave, i.e. the magnetic field is parallel to the boundary and perpendicular to the direction of wave propagation, which, according to figure 4.3, is the $\hat{z}$ direction. In fact, it can be shown that only waves operating in the TM mode can propagate at the interface between the two media, as the reader can verify in reference [1]. Thus, from an initial set of six components, the imposition that the wave is in the TM mode reduces the set to three components: $\mathbf{H}_x$, $\mathbf{E}_y$, and $\mathbf{E}_z$. The amplitudes of these three field components within the same medium are related through Maxwell's equations, while the amplitude of each component in different media is related through boundary conditions. The surface wave propagates in the $\hat{z}$ direction while being attenuated in the $\hat{y}$ direction, so we can write the wave vector as $\mathbf{k}_\mathrm{b} = i\gamma_i\hat{y} + \beta\hat{z}$, where γ_i is the positive extinction coefficient of the ith medium, and β is the propagation constant. Thus, from $\mathbf{k}$, we can write the components of the fields $\mathbf{E}$ and $\mathbf{H}$ in both media:

$$\mathbf{H}_x = \begin{cases} H_{0x}e^{-\gamma_1 y}e^{i\beta z}\hat{x}, & \text{if} \quad y > 0 \\ H'e^{+\gamma_2 y}e^{i\beta z}\hat{x}, & \text{if} \quad y < 0 \end{cases} \tag{4.28}$$

$$\mathbf{E}_y = \begin{cases} E_{0y}e^{-\gamma_1 y}e^{i\beta z}\hat{y}, & \text{if} \quad y > 0 \\ E'e^{+\gamma_2 y}e^{i\beta z}\hat{y}, & \text{if} \quad y < 0 \end{cases} \tag{4.29}$$

$$\mathbf{E}_z = \begin{cases} E_{0z}e^{-\gamma_1 y}e^{i\beta z}\hat{z}, & \text{if} \quad y > 0 \\ E'e^{+\gamma_2 y}e^{i\beta z}\hat{z}, & \text{if} \quad y < 0 \end{cases} \tag{4.30}$$

The Helmholtz equation must be satisfied by each of the field components in each medium. For this to occur, it is necessary that

$$\beta^2 - \gamma_1^2 = \frac{\omega^2}{c^2}\varepsilon_1, \quad \beta^2 - \gamma_2^2 = \frac{\omega^2}{c^2}\varepsilon_2. \tag{4.31}$$

The amplitude of $\mathbf{H}_x$ must satisfy the boundary condition at both media. This implies that the amplitude of $\mathbf{H}_x$ must be continuous at the limit $y \to 0$, which results in $H'_{0x} = H_{0x} = H_0$ in equation (4.28). With $\mathbf{H}_x$ known, we can determine $\mathbf{E}_y$ and $\mathbf{E}_z$ in terms of H_{0x} using Maxwell's equations. For this, we use equation (4.20), considering $\mathbf{k} = i\gamma_i\hat{y} + \beta\hat{z}$. Thus, we obtain

$$E_{0y} = \frac{-\beta}{\omega\varepsilon_0\varepsilon_1}H_0, \quad E'_{0y} = \frac{-\beta}{\omega\varepsilon_0\varepsilon_2}H_0, \quad E_{0z} = \frac{-\gamma_1}{i\omega\varepsilon_0\varepsilon_1}H_0, \quad E'_{0z} = \frac{\gamma_2}{i\omega\varepsilon_0\varepsilon_2}H_0. \tag{4.32}$$

Note that the continuity of $\mathbf{D}_y = \varepsilon\mathbf{E}_y$ is automatically satisfied by $\varepsilon_1 E_{0y}$ and $\varepsilon_2 E'_{0y}$. On the other hand, the continuity of $\mathbf{E}_z$ requires that $E_{0z} = E'_{0z}$, which, according to equation (4.32), leads to

$$-\frac{\gamma_1}{\varepsilon_1} = \frac{\gamma_2}{\varepsilon_2}. \tag{4.33}$$

Since γ_1 and γ_2 are positive, to satisfy equation (4.33), it is necessary that ε_1 or ε_2 be negative. This implies that, considering one of the media as a conductor, the surface wave can only exist at frequencies where $\varepsilon < 0$, which indicates that $\omega < \omega_\mathrm{p}$. Thus, the surface wave can only exist at frequencies where propagation through the medium is prohibited.

The continuity of D_y, or rather, $\varepsilon_0 \varepsilon E_y$, brings a fundamental consequence for the existence of the surface wave. Since ε changes sign from one medium to the other, this implies that E_y must also have opposite signs in the two media, in order for $\varepsilon_0 \varepsilon E_y$ to be continuous. For the value of E_y to change from one medium to the other, it is necessary to have a surface electric charge density in the form of a longitudinal wave that oscillates at the same frequency as $\mathbf{E}_y$, i.e. at frequency ω. Figure 4.4 illustrates the field lines (with the magnetic field perpendicular to the plane of the figure) and the charge distribution on the surface.

What we call SPP is an electromagnetic excitation that occurs at the interface between a metal and a dielectric, resulting from the coupling between the incident electromagnetic wave and the collective oscillations of the free electrons at the metal's surface. The term 'plasmon' refers to these oscillations of surface charge density, while 'polariton' emphasizes the hybrid nature of the excitation, combining characteristics of electromagnetic waves and plasmonic oscillations. A plasmon can be described as a quasiparticle that represents these collective oscillations of free electrons in a conducting material, typically in response to an external electric field. These oscillations are analogous to sound waves in a fluid, where the phonon plays the role of a quasiparticle. In the case of the plasmon, the oscillations occur in the 'gas' of free electrons within the metal. The term 'polariton' implies that this quasiparticle propagates due to the strong coupling with the electromagnetic wave. In other words, the incident electromagnetic wave induces a periodic redistribution of the surface charges in the metal, creating an oscillating charge density that strongly couples with the electromagnetic field, resulting in a surface wave that propagates along the metal–dielectric interface.

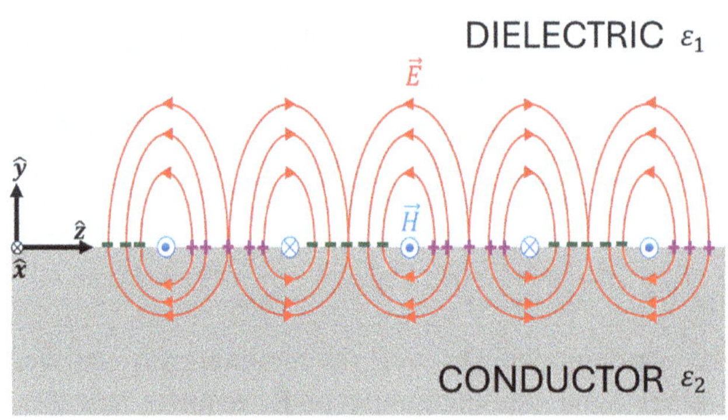

Figure 4.4. Scheme illustrating the SPP field lines at the interface between the dielectric medium and the conducting medium. The electric field lines are shown in red, lying in the plane of the page, with arrows indicating their direction. The magnetic field lines are shown in blue, oriented orthogonally to the page. Plus signs represent positive charges, and minus signs represent negative charges.

This surface wave oscillates at the same frequency as the incident wave, but it has propagation properties distinct from those of the incident wave in each of the two media. From the Helmholtz equation, written for each of the media, we can obtain the propagation parameter β for the surface wave in terms of the dielectric constants ε_1 and ε_2 as

$$\beta = \frac{\omega}{c}\sqrt{\frac{\varepsilon_1\varepsilon_2}{\varepsilon_1 + \varepsilon_2}}. \tag{4.34}$$

From the wave equation, we can define a refractive index n_b and a dielectric constant ε_b for the surface wave,

$$\varepsilon_b = \frac{\varepsilon_1\varepsilon_2}{\varepsilon_1 + \varepsilon_2}; \quad n_b = \sqrt{\varepsilon_b}. \tag{4.35}$$

The subscript 'b' serves as a reminder that this wave propagates at the boundary between the two media. This wave is attenuated as it penetrates both media, and the extinction factors γ_1 and γ_2 can be written as

$$\gamma_1 = \frac{\omega}{c}\sqrt{\frac{-\varepsilon_1^2}{\varepsilon_1 + \varepsilon_2}}; \quad \gamma_2 = \frac{\omega}{c}\sqrt{\frac{-\varepsilon_2^2}{\varepsilon_1 + \varepsilon_2}}. \tag{4.36}$$

Since the surface wave is a propagating wave, it is necessary for the propagation constant β to be positive. This imposes certain restrictions on the frequencies for which it can exist and, consequently, for which it can be excited. In fact, from equation (4.34), we have that for $\beta > 0$, it is necessary that $|\varepsilon_2| > \varepsilon_1$, which imposes a limiting frequency for the existence of the SPP (ω_{sp}) smaller than ω_p. Using equation (4.16), we obtain ω_{sp} as a function of ε_1 as

$$\omega_{sp} = \frac{\omega_p}{\sqrt{1 + \varepsilon_1}}. \tag{4.37}$$

Thus, ω_{sp} is the maximum frequency at which an SPP wave can be excited. For $\omega < \omega_{sp}$, we have $|\varepsilon_2| > \varepsilon_1 \rightarrow \beta > 0$. Note that not every value of $\varepsilon_2 < 0$ allows for the existence of an SPP wave, as shown in figure 4.5(a). For values of $-\varepsilon_1 < \varepsilon_2 < 0$, the incident wave cannot propagate either within or on the surface of the conducting medium, being strongly reflected back into the dielectric medium with a small fraction absorbed through the evanescent fields within real conductors, as mentioned before. Thus, this range constitutes the forbidden band for the existence of SPP waves. For $\omega > \omega_p$, there will be no SPP wave, as the medium behaves like a dielectric, allowing the propagation of electromagnetic waves.

Since $\beta > 0$, we have that the parameters ε_b and n_b are also positive, thus characterizing an actual wave propagating over the dielectric/conductor interface. This same condition ensures that γ_1 and γ_2 are positive, as expected. The surface wave is highly evanescent as it penetrates the two media; however, $\gamma_2 > \gamma_1$, meaning the penetration length in the conducting medium is smaller than in the dielectric

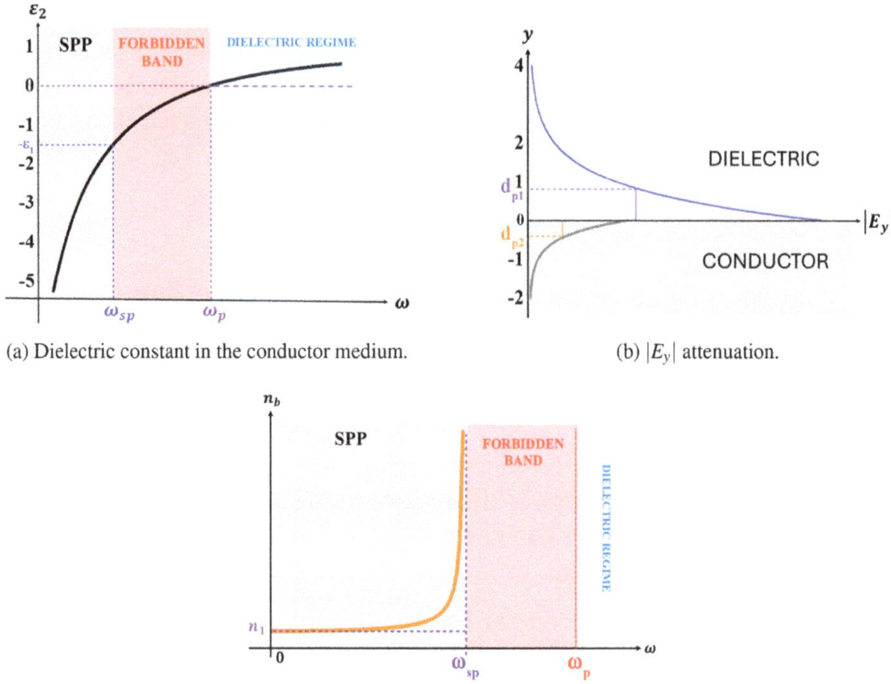

Figure 4.5. (a) Dielectric constant of the conductor as a function of the electromagnetic wave frequency. Each color represents the frequency range where SPP can occur (gray), a forbidden band exists for both SPP and wave propagation inside the conductor (red), and the conductor behaves as a dielectric, allowing electromagnetic wave propagation within its bulk (blue). (b) Decay of the modulus of the SPP electric field $|E_y|$ (x-axis) as a function of the penetration into the two media (y-axis). The interface between the two media is at $y = 0$, with the dielectric medium in $y > 0$ and the conducting medium in $y < 0$. d_{p1} represents the penetration depth in the dielectric medium, and d_{p2} represents the penetration depth in the conducting medium. (c) Refractive index of the SPP wave as a function of the incident wave frequency. The colors indicate the frequency ranges where: SPP can occur (gray), a forbidden band exists (red), and the conductor behaves as a dielectric (blue). n_1 is the refractive index of the dielectric medium.

medium. In both cases, the penetration length is much smaller than the wavelength (in vacuum). Figure 4.5(b) shows the decay of $|E_y|$ in both media as a function of y.

Associated with the surface wave is an average energy flux that can be obtained through the average of the Poynting vector $\langle \mathbf{S} \rangle = \mathrm{Re}(\frac{1}{2}\mathbf{E} \times \mathbf{H}^*)$. Note that, according to equation (4.32), the components H_x and E_y are phase-shifted by $\frac{\pi}{2}$ (since $E_{oz} \propto iH_x$), meaning that no energy flows in the $\hat{y}$ direction, and all energy flows in the $\hat{z}$ direction. This energy is transported through both media, $|y| < d_i$, in the $+z$ direction in the dielectric and in the $-z$ direction in the conductor. The intensity in each medium can be directly obtained from $\langle \mathbf{S} \rangle$,

$$I_1(y) = \frac{\beta}{2\omega\varepsilon_1}|H_0|^2 e^{-2\gamma_1 y}; \quad I_2(y) = \frac{\beta}{2\omega|\varepsilon_2|}|H_0|^2 e^{2\gamma_2 y}, \tag{4.38}$$

where I_1 is the intensity of the surface wave in the dielectric medium and I_2 is the intensity of the surface wave in the conductor. As expected, the intensity decays as it penetrates both media. Since the energy flows in opposite directions in each medium, the net power transported by the SPP wave is $P_N = P_1 - P_2$, where P_i is the power (the area under the I_i curve) in the ith medium. From equation (4.38), we have $P_N \propto \varepsilon_2^2 - \varepsilon_1^2$. Note that for $\omega \approx \omega_{\text{sp}}$, we have $P_N \approx 0$, implying that no net power is transported by the SPP. In fact, as $\omega \to \omega_{\text{sp}}$, the magnitude of the SPP velocity ($v_{\text{spp}} = c/n_{\text{b}}$) tends to zero, since $n_{\text{b}} \to \infty$, as shown in figure 4.5(c).

Our analysis considers the damping of conduction electron oscillations to be negligible, such that ε_2 is written using equation (4.16), making β real. Under this approximation, β diverges as $\omega \to \omega_{\text{sp}}$, leading to the group velocity approaching zero. However, in real conductors, conduction electrons experience damping due to both free electron collisions and interband transitions. Therefore, ε_2 must be complex and described by equation (4.15), resulting in the propagation constant β also being complex. This causes SPP waves to be attenuated as they propagate. The characteristic attenuation length, also known as the propagation length L, is defined as $L = (2\,\text{Im}[\beta])^{-1}$, which typically ranges from 10 to 100 μm in the visible spectrum.

With the inclusion of damping in conduction electrons, Re[β] attains a finite (and maximum) value at $\omega = \omega_{\text{sp}}$. Another change that occurs in real conductors compared to the ideal conductor limit is the decay of the evanescent wave, which is expressed as $e^{-|\gamma_1|y}$ and $e^{-|\gamma_2|y}$ for the dielectric and conductor media, respectively. The penetration depth in these media is defined as $d_{\text{p}} = 1/|\gamma|$. For instance, at the silver/air interface with $\lambda_0 = 450$ nm, we have $L \approx 16$ μm, $d_{p1} \approx 180$ nm, and $d_{p2} \approx 20$ nm. At $\lambda_0 = 1500$ nm, these values change to $L \approx 1080$ μm, $d_{p1} \approx 2.6$ μm, while $d_{p2} \approx 20$ nm remains nearly unchanged.

Notably, in the dielectric medium, it is possible to confine the field below the diffraction limit as $\omega \to \omega_{\text{sp}}$. For $\lambda_0 = 450$ nm, the field is confined within $d_{\text{p}} = 180$ nm, which is less than half the wavelength. This high confinement results in an increase in the local field intensity, which is one of the main features of plasmonics, enabling numerous applications in optical sensors and, as we will see later, enhancing the efficiency of optical trapping. The reason for this is that the SPP wavelength, $\lambda_{\text{spp}} = \lambda_0/n_{\text{b}}$, is always smaller than λ_0, with its minimum value at ω_{sp} where $n_{\text{b}} \to \infty$ for ideal conductors (as shown in figure 4.5(c)) and reaching its maximum value in real conductors. Indeed, since $n_{\text{b}} \geqslant n_1$, where n_1 is the refractive index of the dielectric medium, it follows that $\lambda_{\text{spp}} \leqslant \lambda_1$, with $\lambda_1 = \lambda_0/n_1$ representing the wavelength in the dielectric medium. This implies that the field can be confined below the diffraction limit regardless of the dielectric medium used to excite the SPP. The confinement of the field below the diffraction limit simultaneously leads to the confinement of energy. For readers interested in understanding how energy is stored below the diffraction limit in SPPs, we recommend referring to section 2.4 of [1].

There is another condition that must be met for the SPP wave to be excited by an electromagnetic wave, in addition to $\omega < \omega_{\text{sp}}$. Considering that the incidence plane is the yz plane, the wave vector of the incident wave is written as $\mathbf{k}_i = k_i\hat{y} + k_{iz}\hat{z}$, and, as we have shown, the wave vector of the SPP wave can be written as $\mathbf{k}_\text{b} = i\gamma\hat{y} + \beta\hat{z}$, with the term $i\gamma$ representing the evanescent nature of the SPP wave in both media. For the SPP wave to be excited, a match must occur between the real part of $\mathbf{k}_\text{b}$ and $\mathbf{k}_i$. In other words, the projection of the wave vector along the propagation direction must be equal to the propagation constant of the SPP, $\beta = k_{iz}$.

This requirement adds an additional challenge to the generation of SPPs, as the SPP dispersion curve (solid gray line) lies to the right of both the light line (dashed orange line) and the light dispersion in the dielectric (dashed blue line), as shown in figure 4.6(a). Consequently, it is not possible to excite SPPs through direct illumination on the dielectric/conductor interface, since $\beta > k_i$ for any $\omega \neq 0$, where k_i is the wave vector magnitude of light on the dielectric side of the interface. Given that $k_{iz} = k_i \sin\theta$, where θ is the angle of incidence of the light relative to the normal, it becomes evident that the coupling condition will not be satisfied. Therefore, certain techniques must be employed to achieve SPP excitation, such as using charged particle beams [5–7], grating coupling [8, 9], among others [10, 11].

In addition to these techniques, the coupling condition can be achieved in a three-layer system, which involves placing a thin conductive film between two dielectrics with different dielectric constants. Typically, one of the dielectric media is a prism with a dielectric constant ε_p higher than the dielectric constant of the other medium, ε_1 (figure 4.6(b)). A beam reflected at the interface between the prism and the metal will have a propagation direction momentum component $K_{iz} = k_0\sqrt{\varepsilon_\text{p}}\sin\theta$, which can be sufficient to excite the SPP at the metal/dielectric

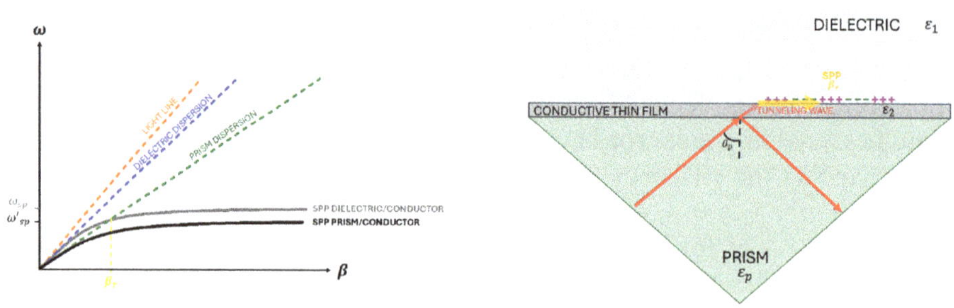

(a) Dispersion relation. (b) SPP Generation.

Figure 4.6. (a) Dispersion relation: light line (dashed orange line), dielectric medium (dashed blue line), prism (dashed green line), SPP created at the interface between the dielectric medium and the conductor (solid gray line), SPP created at the interface between the prism and the conductor (solid black line). ω_SP and ω'_SP indicate the maximum frequency at which the SPP can be excited at the dielectric/conductor and prism/conductor interfaces, respectively. (b) Generation of an SPP wave using a prism coupler. A beam is incident on the prism at an angle θ_p, and the incident wave tunnels through the conducting film, reaching the conductor/dielectric interface. When $\theta_\text{p} = \theta_\text{r}$ such that $k_{iz} = \beta = \beta_r$, an SPP wave is generated at the conductor/dielectric interface. Note that it is not possible to generate an SPP wave at the prism/conductor interface.

interface. This is possible because the electromagnetic wave from the prism has a dispersion curve (dashed green line, figure 4.6(a) that intersects the SPP dispersion curve created at the conductor/dielectric interface (solid gray line).

It is important to note that it is not possible to create an SPP at the prism/conductor interface because the dispersion curve for this case lies to the right of the light dispersion curve in the prism (solid black line, figure 4.6(a). The prism shape of the second dielectric is advantageous because, by varying the angle of incidence θ_p of the electromagnetic wave, we can tune the value of $k_{iz} = \beta$. This condition is met at a specific angle θ_r, where $\beta = \sqrt{\varepsilon_p} k_0 \sin \theta_r$. It is straightforward to see that this condition is satisfied when the angle of incidence θ_r is

$$\sin \theta_r = \sqrt{\frac{\varepsilon_l}{\varepsilon_p} \frac{1 - \omega^2/\omega_p^2}{1 - (1 + \varepsilon_l)\omega^2/\omega_p^2}}. \tag{4.39}$$

For the wave coming from the prism to excite the SPP at the conductor/dielectric interface, it must tunnel through the conductor, which is why a thin film is necessary. When the phase-matching condition is met, optical power is transferred to generate the SPP through a form of frustrated total internal reflection (FTIR), causing the power of the wave reflected at the prism/conductor interface to decrease significantly. The reflection intensity curve resembles a resonance curve, with a minimum at the coupling angle θ_r. This sensitivity of the reflected wave intensity at the prism/conductor interface to the coupling angle θ_r forms the basis of the surface plasmon resonance (SPR) spectroscopy technique.

SPR is a technique that measures variations in the refractive index near a metallic surface, typically gold or silver, through SPP excitation. When light is incident at the angle θ_r, exciting the SPP, it leads to a reduction in the intensity of the reflected light. Changes in reflected intensity, caused by variations in the refractive index near the surface, can be monitored to detect molecular interactions, such as protein binding [12] or concentration changes [13], making SPR a valuable tool in biology, chemistry, and biomolecule detection [1, 14].

4.3 SPP excitation using highly focused optical beams

The prism configuration discussed in the previous section is often referred to as the Kretschmann method [15], which, together with the Otto configuration [16], constitutes one of the most common geometries for exciting SPPs through the three-layer method. However, with the advancements in plasmonics, new methods for exciting SPPs have emerged, one of particular interest in the field of optical tweezers (OT). As a variation of the Kretschmann method, a high numerical aperture (NA) objective lens is used instead of a prism to excite SPPs. One of the pioneering works exploring this setup was carried out by Bouhelier and Wiederrecht [10], whose original setup is shown in figure 4.7.

An oil immersion objective lens is placed in contact with the glass substrate (refractive index n_g) through an immersion oil layer with a matching index. On the

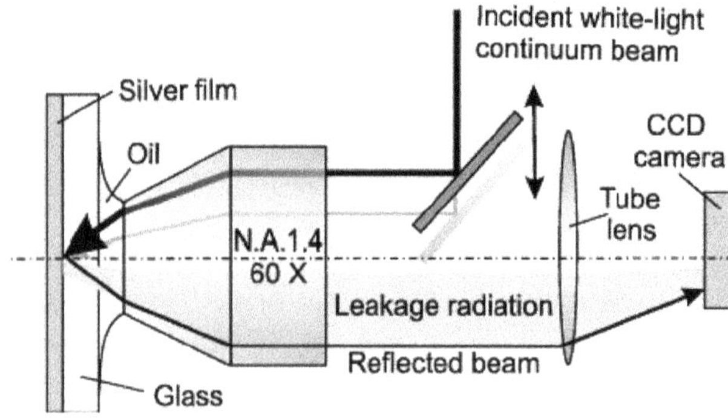

Figure 4.7. Schematic of the excitation of a white-light continuum of SPPs and their observation via detection of the leakage radiation using an index-matched oil immersion lens. Reprinted with permission from [10], copyright 2005 The Optical Society of America.

opposite side of the glass substrate, a silver film with a thickness of approximately 45 ± 5 nm is deposited by thermal evaporation, in contact with air. Before delving into the role of the objective lens and the importance of its NA and oil immersion, it is crucial to understand a key concept: leakage radiation (LR).

Leakage radiation refers to the radiation emitted when SPPs propagate along a metal–dielectric interface, and part of the energy of the SPPs couples out of the interface, transforming into radiation that can be detected in the far field. In Bouhelier's work, LR is used to visualize and measure the intensity distribution of SPPs. When SPPs are excited at the air–silver interface, they propagate along this interface, and part of their energy couples into the opposite interface (metal–glass), where it transforms into detectable radiation. This leakage radiation enables the direct observation of SPP propagation and the measurement of their propagation length, as it can be collected by the objective and analyzed by a CCD camera.

The high NA of the objective ensures a large angular dispersion of the focused excitation beam, as well as focusing the beam into a spot smaller than the SPP propagation length (L), which is essential for visualizing SPP propagation. The angular dispersion ensures the availability of various wave vectors $\mathbf{k}_i$, such that the coupling condition $k_{iz} = \beta$ is satisfied for several wavelengths. Indeed, the high NA of the objective allows a white light beam (continuum of wavelengths) to be focused on the interface. As a result, practically all wavelengths of the white light can excite SPPs, forming a rainbow jet of SPPs.

The phase-matching between the incident wave and the SPP occurs at specific angles (θ_{spp}) for each wavelength. Generally, $\theta_{\text{spp}}^{\lambda} = \arcsin(\beta_\lambda / n_g k_0)$, where $n_g = \sqrt{\varepsilon_g}$ is the refractive index of the glass for the wavelength λ. Notably, $\theta_{\text{spp}}^{\lambda} > \theta_c$, where θ_c is the critical angle for total internal reflection at the glass/air interface. To collect LR for SPP visualization, it is necessary for LR to be collected by the objective. If the contact interface of the objective with the glass substrate contains air (or another fluid with $n < n_g$), such that $\theta^{\lambda}_{spp} > \theta_c$, LR will undergo total internal reflection at the glass/air interface and will not be collected by the objective. To prevent this, an oil immersion objective is used to match the refractive index of the oil with that of the glass, ensuring that total internal reflection does not occur.

Note that not all angles from the objective contribute to SPP excitation. Within the visible spectrum, the dispersion of θ^{λ}_{spp} values is small, and all light from other angles is either transmitted through the metal/air interface or reflected back into the objective. To optimize the setup, instead of fully illuminating the back aperture of the objective, it is preferable to collimate the light into a small beam and direct it into the objective to optimize the angle θ^{λ}_{spp}, as shown in figure 4.7.

When SPPs are excited with white light, in real conductors, each wavelength generates an SPP at a specific coupling angle. For each SPP, an LR is emitted at the metal/glass interface with the same frequency, distributed along the in-plane wave vector corresponding to the projection of the excitation vector on the film surface. As a result, a rainbow jet near the excitation point can be observed, as shown in figure 4.8(a). The bright spot is the reflection of illumination at the metal/glass interface. Note that the blue light is confined closer to the spot, while the red light propagates over a greater distance. This spectral intensity distribution can be understood as a result of the β dispersion curve (4.6(a)). For shorter wavelengths (such as blue), the SPP dispersion curve flattens asymptotically towards ω_{sp}, resulting in lower group velocities and higher intrinsic losses due to metal absorption. Thus, the blue part of the spectrum tends to remain more confined to the excitation point, while longer wavelengths propagate further along the film surface. In other words, shorter wavelengths have a smaller propagation length than longer wavelengths.

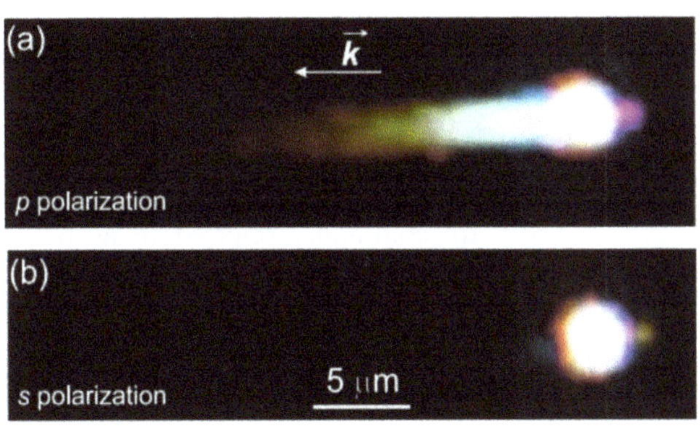

Figure 4.8. (a) Leakage radiation intensity distribution for a TM polarized white-light continuum excitation beam, showing SPPs propagating away from the excitation spot. (b) No SPP excitation is observed for TE polarization. Reprinted with permission from [10], copyright 2005 The Optical Society of America.

The reader should recall that SPPs operate in the TM mode (also known as p-polarization). For coupling between the excitation beam and the SPP to occur, the excitation beam must also have p-polarization. When the beam polarization is switched to s-polarization, no SPPs can be excited, and consequently, no LR can be observed, as shown in figure 4.8(b).

The SPP generated from the focusing of an objective lens is an essential component in the construction of so-called plasmonic tweezers, which use the electric field originating from the SPPs (or more precisely, from the corresponding LRs) to create the optical trap. As we will see, these fields are characterized by being narrower and more intense than the excitation fields, which optimizes the trapping process and allows for the confinement of different types of materials, such as micrometer-sized metallic particles, which are challenging to trap using traditional OT.

4.4 Localized surface plasmons and plasmonic particles

In the previous sections, we discussed how SPPs are propagating electromagnetic waves coupled to the free electrons of a conductor at the interface with a dielectric. In addition to this coupling mode, another important mechanism emerges when the conductor dimensions are on the order of (or smaller than) the wavelength of the incident electromagnetic wave. This new coupling mode, known as *LSPs*, consists of non-propagating excitations (hence the term 'localized') of free electrons in metallic nanostructures coupled to an external electromagnetic field.

LSPs naturally arise from the scattering of an oscillating electromagnetic field by a conducting nanostructure. The interaction between the external field and the free electrons induces an oscillating dipole in the nanostructure, whose resonance amplifies both the internal electromagnetic fields and the near-field outside the structure. The curvature of the nanostructure surface plays a crucial role, as it imposes a restoring force on the free electrons, determining the resonance frequency. This frequency depends on the nanostructure size, shape, and the dielectric properties of both the metal and the surrounding medium.

A key advantage of LSPs, in contrast to SPPs, is that they can be directly excited by light without requiring phase-matching. This eliminates the need for sophisticated coupling geometries, as discussed in section 4.2, allowing for a more straightforward and efficient excitation process.

Within the OT community, conducting nanostructures (typically spheres or spheroids) that exhibit LSPs are commonly referred to as plasmonic particles. These particles demonstrate significantly higher optical trapping efficiency compared to dielectric particles under the same conditions. This enhancement primarily results from the amplification of electromagnetic fields at the LSP resonance, which increases the optical forces.

Although metallic nanoparticles exhibit strong trapping efficiency, their behavior changes for larger particles (mesoscopic and microscopic scales). For particles with dimensions comparable to or larger than the wavelength, the quasi-static approximation is no longer valid, and multipolar modes become significant. Additionally, scattering and absorption losses increase considerably, reducing trapping efficiency. Consequently, the advantages observed in plasmonic nanoparticles do not extend directly to larger particles.

In the following sections, we will explore in detail the fundamental aspects of plasmonics, focusing on the excitation and properties of LSPs in metallic nanostructures. We will discuss how factors such as nanoparticle size, shape, and surrounding dielectric environment influence plasmonic resonance. Additionally, we will investigate how LSPs govern the interaction between light and conducting nanoparticles, laying the theoretical foundation necessary to understand their role in optical trapping. This discussion will provide the essential background for the next section, where we will analyze how LSPs enhance optical forces in the context of OT.

4.4.1 LSP in nanostructures: the quasi-static approximation

Although all the mathematical framework in this section has already been discussed in section 2.1, there are physical subtleties that we need to understand in order to comprehend the interaction between conductive nanoparticles and light. Our starting point is the quasi-static approximation, justified by the relationship between the particle size and the wavelength of light in the surrounding medium. There is no new justification here: since the particle radius (a) is much smaller than the wavelength, the phase of the time-harmonic electromagnetic wave can be considered practically constant over the particle volume. Thus, the spatial distribution of the electric field can be calculated by assuming a simplified problem in which the particle is subjected to a uniform electrostatic field, as discussed in section 2.1. Therefore, our first task is to determine the spatial distribution of the electric field $\mathbf{E}$ throughout the space, i.e. inside the particle ($\mathbf{E}_{\text{in}}$) and outside it ($\mathbf{E}_{\text{out}}$), considering that the particle, assumed to be spherical for simplicity, interacts with a constant external field $\mathbf{E}_0 = E_0\hat{z}$.

The first subtle point to be understood is the justification for using Laplace's equation, given that the conductive particle has a free electron density on the order of 10^{22}–10^{23} cm^{-3}. At first glance, one might consider that the problem should be treated by assuming a volumetric charge density ρ interacting with an external field $\mathbf{E}_0$, obtaining $\mathbf{E}$ from Poisson's equation. However, the answer is **no**: the use of Laplace's equation is correct and is justified precisely by the high density of free electrons.

Remember that the plasma frequency ω_{p} is proportional to the free electron density, so that, for metals, ω_{p} is typically in the ultraviolet spectrum. Since we are interested in interactions occurring in the visible to infrared spectral region, we have $\omega_{\text{p}} \gg \omega$. This implies that the response time of the charges (on the order of $1/\omega_{\text{p}}$) is much shorter than the oscillation period of the external electric field $\mathbf{E}_0$ (on the order of $1/\omega$), causing the charges to respond almost 'instantaneously' to the presence of $\mathbf{E}_0$.

In practice, the incident electric field $\mathbf{E}_0$ induces a displacement of the free electrons in the conductor, accumulating negative charge in a specific region of the particle surface. Consequently, the region from which the electrons have migrated exhibits an excess of positive charge (due to the positive ions in the crystal lattice). Equilibrium is reached when these charge distributions adjust to minimize the electrostatic potential between them. This adjustment results in the charge being distributed exclusively on the surface of the conductor. Since this equilibrium is achieved in a time much shorter than the oscillation period of the external field, we can assume that the charge distribution is always in equilibrium. Thus, as there is no free charge inside the conductive particle, Laplace's equation can be employed to describe the electrostatic potential, and the surface charge can be computed as a boundary condition, exactly as we did in section 2.1 (boundary condition 2).

Indeed, the same set of boundary conditions presented in section 2.1 is used here, leading to the same mathematical solutions. However, although the equations are identical, the physical interpretation imposed by the conductive nature of the particle is different. Let us begin by analyzing the internal field of the conducting sphere, which, according to equation (2.15), can be written as:

$$\mathbf{E}_{\text{in}} = \frac{3\varepsilon_{\text{m}}}{\varepsilon_{\text{c}} + 2\varepsilon_{\text{m}}}\mathbf{E}_0 \tag{4.40}$$

where ε_c (provided by equation (4.16)) is the dielectric constant of the conductor, and ε_m is the dielectric constant of the surrounding medium (assumed to be a non-absorbing dielectric). In the limit of a perfect conductor ($\varepsilon_c \to -\infty$), we obtain $\mathbf{E}_{\text{in}} = 0$, as expected. However, for real conductors, where ε_c is finite, the internal field $\mathbf{E}_{\text{in}}$ not only exists but also plays a crucial role in LSP resonance.

To understand the nature of $\mathbf{E}_{\text{in}}$, it is essential to highlight that, unlike dielectrics, where the internal field originates from the material's polarization, in conductors it emerges from the collective response of free electrons to the applied external field. When an external electric field $\mathbf{E}_0$ is applied, the free electrons in the nanoparticle shift, creating a charge separation at the particle surface. This redistribution generates an internal restoring field that opposes the charge displacement, establishing a dynamic equilibrium. Since the response of free electrons occurs on a timescale much shorter than the oscillation period of the external field, this charge redistribution can be considered practically instantaneous. The coherent oscillation of these surface charges gives rise to a natural resonance frequency, denoted as ω_{LSP}, at which LSP resonance occurs.

Neglecting the effect of bound electrons (which will be addressed in chapter 6 and is only relevant for metals at high frequencies), there is no charge separation inside the particle, as occurs in dielectrics. This means that there is no volumetric dipole moment inside the conducting sphere, reinforcing the distinction between the electric field in dielectrics and conductors. However, for points external to the sphere, the redistribution of surface charges generates an electric field that, to first order in the multipole expansion, behaves like the field of an induced electric dipole. Thus, we can define a dipole moment equivalent to the one introduced in equation (2.16) for dielectrics, but with a distinct physical origin. This implies that, unlike in dielectrics, where material polarization defines a volumetric dipole moment, in conductors this dipole moment is generated exclusively by the redistribution of surface charges. Thus, the dipole moment $\mathbf{p}$ is

$$\mathbf{p} = 4\pi\varepsilon_0\varepsilon_m a^3\left(\frac{\varepsilon_c - \varepsilon_m}{\varepsilon_c + 2\varepsilon_m}\right)\mathbf{E}_0. \tag{4.41}$$

From this, we can define the polarizability of the conducting particle,

$$\alpha = 4\pi a^3\frac{\varepsilon_c - \varepsilon_m}{\varepsilon_c + 2\varepsilon_m}. \tag{4.42}$$

Unlike in dielectrics, where polarization occurs due to the separation of bound charges within the material, in conductors the response is dominated by the collective motion of free electrons and the consequent redistribution of charge on the surface. Since free electrons can move more easily in response to an external electric field, the charge separation in conducting particles is more significant, resulting in a stronger polarization. Moreover, the SPR mechanism further amplifies this response, leading to a significant increase in polarizability α. The electric field outside the particle ($\mathbf{E}_{out}$) is, in turn, the result of the superposition of the incident field $\mathbf{E}_0$ and the dipole field induced by the conducting particle $\mathbf{E}_{dip}$, which, according to equation (2.18), can be written as

$$\mathbf{E}_{out} = \mathbf{E}_0 + \frac{3\hat{n}(\hat{n}\cdot\mathbf{p}) - \mathbf{p}}{4\pi\varepsilon_0\varepsilon_m r^3}, \tag{4.43}$$

where the first term represents the incident field, and the second term corresponds to the dipole field. Here, $\hat{n}$ is the unit vector in the direction of the point where the field value is to be calculated. The induced surface charge density can be obtained from the discontinuity of the normal component of the electric field $\mathbf{E}$. Thus, using equation (2.21), we have

$$\sigma = \varepsilon_0\varepsilon_m(\mathbf{E}_{out} - \mathbf{E}_{in})\cdot\hat{r} = 3\varepsilon_0\varepsilon_m\left(\frac{\varepsilon_c - \varepsilon_m}{\varepsilon_c + 2\varepsilon_m}\right)E_0\cos\theta, \tag{4.44}$$

where $\hat{r}$ is the normal direction to the sphere. Note that equation (4.44) is similar to its dielectric counterpart but with ε_c replacing ε_p. However, the fact that ε_c can take negative values leads to substantially different physical effects. If we use equation (4.16) to describe ε_c, we find that as $|\varepsilon_c| \to 2\varepsilon_m$, $\sigma \to \infty$, meaning that the induced surface charge density on the sphere would tend to an infinite value. This is, of course, a non-physical condition.

The issue is resolved by including the damping of free electrons and describing ε_c using equation (4.15), ensuring that σ reaches a maximum (but finite) value when $\mathrm{Re}(\varepsilon_c) = -2\varepsilon_m$. This condition, known as the Fröhlich condition, describes the resonance of surface plasmons in metallic nanoparticles. Notably, it is not only the charge density that exhibits this resonance condition but also the polarizability and the internal and external electric fields of the conducting sphere. This enhancement of the electric field at plasmonic resonance underlies many applications of metallic nanoparticles in sensors and optical devices.

Thus, within the quasi-static approximation, a conducting sphere behaves as an induced dipole interacting with an electrostatic field. Returning to the initial problem and introducing the temporal variation in $\mathbf{E}_0$, treating it as a plane wave $\mathbf{E}_0(\mathbf{r}, t) = \mathbf{E}_0(\mathbf{r})e^{i\omega t}$, the representation as an ideal dipole remains appropriate as long as the quasi-static approximation holds, i.e. the electric field remains uniform over the particle volume (a condition satisfied for the plane wave when $a \ll \lambda$). This allows for the consideration of time-varying fields while neglecting spatial retardation effects within the conducting sphere, since the spatial variation of $\mathbf{E}_0(\mathbf{r})$ remains negligible inside the particle.

Illumination with a plane wave induces a time-dependent dipole moment that oscillates at the same frequency ω, given by

$$\mathbf{p}(\mathbf{r}, t) = \varepsilon_0 \varepsilon_m \alpha \mathbf{E}_0(\mathbf{r}) e^{i\omega t}, \tag{4.45}$$

where α is the same electrostatic polarizability from equation (4.42). The temporal variation of the dipole moment results from the oscillation induced by the plane wave in the surface charge density. As the amplitude of the incident electric field oscillates, the charges induced on the conductor's surface also oscillate at the same frequency. Unlike in dielectrics, where this oscillation is purely forced, in conductors it can become resonant, depending on the frequency of $\mathbf{E}_0$. The resonance occurs when $\omega = \omega_{\mathrm{LSP}}$, arising from the coupling between the incident electromagnetic field and the coherent oscillations of electrons at the metal–dielectric interface.

In other words, when the external field oscillates at the resonant frequency, the free charges on the conductor surface enter a state of collective oscillation with an amplitude much greater than at other frequencies. This drastic increase in the amplitude of the field near the particle leads to a localized enhancement of the electric field. Notably, this resonant frequency maximizes the polarizability of the conducting particle and is determined by the Fröhlich condition. Thus, we can define ω_{LSP} as

$$\mathrm{Re}\{\varepsilon_c(\omega_{\mathrm{LSP}})\} + 2\varepsilon_m = 0, \tag{4.46}$$

$$\omega_{\mathrm{LSP}} = \frac{\omega_p}{\sqrt{1 + 2\varepsilon_m}}. \tag{4.47}$$

It is important to note that the frequency, ω_{LSP}, is distinct from both the SPP frequency, ω_{SPP}, and the plasma frequency, ω_p. Physically, ω_{LSP} can be interpreted as the natural oscillation frequency resulting from the Coulombic force associated with the separation of surface charges. Indeed, the ratio $\omega/\omega_{\mathrm{LSP}}$ determines the nature of the internal field $\mathbf{E}_{\mathrm{in}}$ and, consequently, the distribution of surface charges in the conducting sphere.

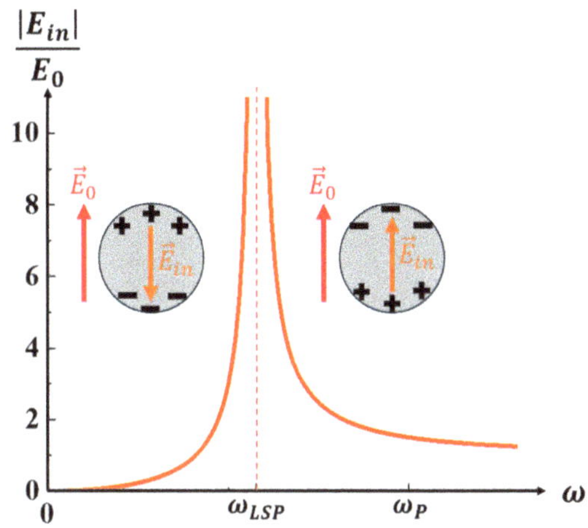

Figure 4.9. Diagram illustrating the distribution of the induced surface charge density on the particle as a function of ω. Below the resonance ($\omega < \omega_{\mathrm{LSP}}$), the internal field is opposite to the incident field (out of phase), whereas above the resonance ($\omega > \omega_{\mathrm{LSP}}$), they have the same direction (in phase).

This relationship can be visualized in figure 4.9. For frequencies below resonance ($\omega < \omega_{\mathrm{LSP}}$), $\mathbf{E}_{\mathrm{in}}$ is out of phase with $\mathbf{E}_0$, meaning there is a phase shift close to π. This phase shift results in a surface charge distribution analogous to that observed in dielectric particles when $\varepsilon_{\mathrm{p}} > \varepsilon_{\mathrm{m}}$ (figure 2.3(a)). Conversely, for frequencies above ω_{LSP}, the induction of surface charges in the conducting particle resembles that observed in dielectric particles with $\varepsilon_{\mathrm{p}} < \varepsilon_{\mathrm{m}}$ (figure 2.3(b)).

It is insightful to compare the behavior of the internal field $\mathbf{E}_{\mathrm{in}}$ in conducting particles (figure 4.9) with its behavior in dielectric particles (figure 2.4). This comparison highlights the effect of LSP resonance, which significantly enhances the magnitude of the internal field relative to $\mathbf{E}_0$.

Just as in dielectrics, the oscillatory dynamics of the induced surface charges give rise to a polarization current density $\mathbf{J}$. This current density acts as the source of the external field generated by the conducting sphere in electrodynamics. Furthermore, the presence of $\mathbf{J}$ leads to the emergence of a magnetic field $\mathbf{H}$ alongside the electric field $\mathbf{E}$.

The derivation of these fields follows the same steps as in section 2.1: first, the vector potential $\mathbf{A}$ is constructed, and then an expansion in powers of k (multipole expansion) is performed. Taking the first nonzero term in the expansion, the dipole term, the magnetic and electric fields can be obtained using equations (2.27) and (2.28), respectively.

However, unlike the case of dielectrics, where our focus was on the fields in the far zone (radiation zone), in conductors, the near-zone fields play a crucial role in plasmonics. This is because the confinement and enhancement effects of the electromagnetic field near the conductor surface are fundamental for the excitation of surface plasmons and their applications in spectroscopy and light confinement in conducting nanostructures.

Considering only the dipole term of the vector potential, we have

$$\mathbf{A} = -\frac{i\mu_0\omega}{4\pi}\frac{e^{ikr}}{r}\mathbf{p}, \tag{4.48}$$

so that the total field produced by the sphere is given by

$$\mathbf{H} = \frac{ck^2}{4\pi}(\mathbf{r} \times \mathbf{p})\frac{e^{ikr}}{r}\left(1 - \frac{1}{ikr}\right), \tag{4.49}$$

$$\mathbf{E} = \frac{1}{4\pi\varepsilon_0\varepsilon_\mathrm{m}}\left\{k^2(\mathbf{r} \times \mathbf{p}) \times \mathbf{r}\frac{e^{ikr}}{r} + [3\mathbf{r}(\mathbf{r} \cdot \mathbf{p}) - \mathbf{p}]\left(\frac{1}{r^3} - \frac{ik}{r^2}\right)e^{ikr}\right\}, \tag{4.50}$$

where $k = \frac{2\pi n}{\lambda_0}$ is the wavenumber in the surrounding medium, and $\mathbf{r}$ is the unit vector in the direction of the point where the fields are to be evaluated. Equations (4.49) and (4.50) describe the magnetic and electric fields produced by the conducting sphere at any position outside the sphere, covering the near, intermediate, and far zones.

In the near zone, where $kr \ll 1$, the fields $\mathbf{H}$ and $\mathbf{E}$ can be written as

$$\mathbf{H}_\mathrm{near} = \frac{i\omega}{4\pi}(\mathbf{r} \times \mathbf{p})\frac{1}{r^2}, \tag{4.51}$$

$$\mathbf{E}_\mathrm{near} = \frac{3\mathbf{r}(\mathbf{r} \cdot \mathbf{p}) - \mathbf{p}}{4\pi\varepsilon_0\varepsilon_\mathrm{m}r^3} \tag{4.52}$$

Note that, in the near zone, the electrostatic result for the electric field ($\mathbf{E}_{dip}$) is recovered. Moreover, the fields are predominantly electric. To make this clearer, we compute the ratio between the magnitude of the magnetic field and the electric field,

$$\frac{|\mathbf{H}_\mathrm{near}|}{|\mathbf{E}_\mathrm{near}|} \approx \varepsilon_0\varepsilon_\mathrm{m}\omega r \approx \sqrt{\frac{\varepsilon_0}{\mu_0}}(kr). \tag{4.53}$$

Thus, for static fields ($kr \to 0$), the magnetic field vanishes, and in the near zone, we are left with only the static electric field (hence, this region is also called the electrostatic zone). Due to the LSP resonance, the electric field is amplified, becoming significantly larger than the incident field $\mathbf{E}_0$. The fact that this amplification is confined to the regions near the conducting particle gives the term 'localized' to this type of plasmonic excitation.

This localization is a consequence of the non-propagating nature of the conduction electron oscillations, which are confined by the curved surface of the nanoparticle. As a result, the amplified electric field decays rapidly with distance from the nanoparticle surface, remaining intense only in the near zone. Therefore, the term 'localized' in LSP refers to the characteristic that the electric field enhancement is spatially restricted to the immediate vicinity of the nanoparticle, in contrast to propagating surface plasmons, which spread along an interface. This localization property is crucial for various applications in optical devices and

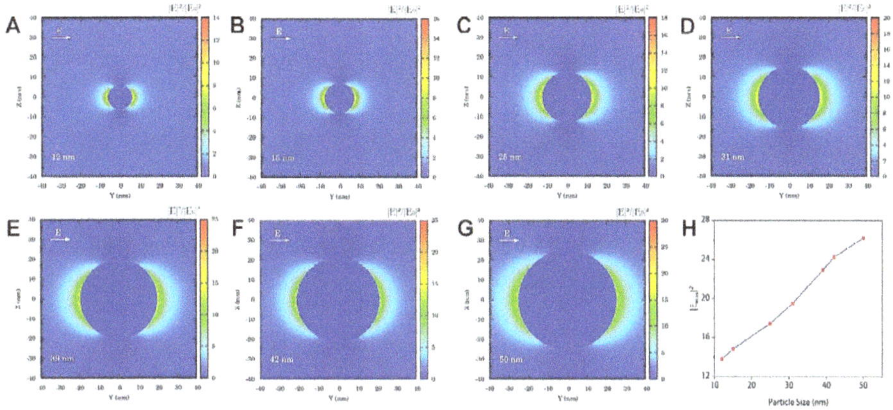

Figure 4.10. Electric field enhancement contours for AgNPs as a function of particle size. The AgNPs were 12 (A), 15 (B), 25 (C), 31 (D), 39 (E), 42 (F), and 50 nm (G) in diameter. The wavelength of the incoming radiation was 633 nm in all cases. (H) $|E_{\max}|/|E|$ values calculated for the AgNPs as a function of size. [18] John Wiley & Sons. Copyright 2018 Wiley-VCH Verlag GmbH & Co. KGaA, Weinheim.

sensors, where near-field enhancement can be exploited to increase sensitivity and efficiency.

Within the size range where the quasi-static approximation holds, both the polarizability and the electric field enhancement are proportional to the particle size. Figure 4.10, adapted from reference [18], illustrates the contours of electric field enhancement (*E-field*) for silver nanoparticles (AgNPs) of different sizes, ranging from 12 to 50 nm in diameter, under incident radiation with a wavelength of 633 nm.

Images (A–G) show that the electric field enhancement ($[E_{\max}]/[E_0]$) is concentrated along the polarization axis of the incident electromagnetic wave (y-axis) and increases with the nanoparticle size. Specifically, the values of $[E_{\max}]/[E_0]$ range from approximately 14 for nanoparticles of 12 nm up to about 27 for nanoparticles of 50 nm, as shown in the graph (H).

Going to the opposite limit, the radiation zone, where $kr \gg 1$, the fields **H** and **E** (equations (4.49) and (4.50), respectively) can be written as

$$\mathbf{H} = \frac{ck^2}{4\pi}(\mathbf{r} \times \mathbf{p})\frac{e^{ikr}}{r}, \tag{4.54}$$

$$\mathbf{E} = \sqrt{\frac{\mu_0}{\varepsilon_0 \varepsilon_m}}\,\mathbf{H}_{\text{rad}} \times \mathbf{r}. \tag{4.55}$$

These expressions are equivalent to equations (2.36) and (2.37) for dielectric particles.

Note that the dependence of $\mathbf{p}$ in $\mathbf{H}_{\text{rad}}$ and $\mathbf{E}_{\text{rad}}$ makes the conducting particle both a resonant scatterer and absorber. This becomes evident when we write the extinction and scattering cross-sections for a conducting nanosphere,

$$\sigma_{\text{ext}} = k\ \text{Im}\,\{\alpha_{\text{rad}}\}, \tag{4.56}$$

$$\sigma_{\text{sct}} = \frac{k^4}{6\pi} |\alpha_{\text{rad}}|^2. \tag{4.57}$$

Again, k is the wavenumber in the surrounding medium, and α_{rad} is the polarizability corrected by the radiation reaction. The definition of σ_{ext} and σ_{sct} follows the same approach as in section 2.1, meaning that σ_{ext} is the ratio of the power extinguished from the incident wave to the incident wave intensity, while σ_{sct} is the ratio of the power scattered by the conducting nanosphere to the incident wave intensity. Additionally, the relation $\sigma_{\text{ext}} = \sigma_{\text{sct}} + \sigma_{\text{abs}}$ holds, where σ_{abs} represents the absorption cross-section.

Since the scattering cross-section depends on the sixth power of the nanosphere radius, for small nanospheres, the extinction cross-section is dominated by absorption. For this reason, many textbooks define equation (4.56) as the absorption cross-section. However, as shown in figure 4.11, which presents the calculated optical efficiency spectra ($\sigma/\pi a^2$) for extinction (figure 4.11(A)), absorption (figure 4.11(B)), and scattering (figure 4.11(C)) of AgNPs with varying sizes, only for small NPs do we actually have $\sigma_{\text{abs}} = \sigma_{\text{ext}}$. In general, the following relation holds:

$$\sigma_{\text{abs}} = \sigma_{\text{ext}} - \sigma_{\text{sct}}. \tag{4.58}$$

Now let us analyze the spectra from figure 4.11 and understand what we mean by a resonant (absorptive) scatterer. In section 2.1, we discussed that Rayleigh scattering is characterized by the $k^4 \propto \omega^4$ dependence, implying that higher frequencies (shorter wavelengths) are more efficiently scattered by a dielectric nanosphere than lower frequencies. However, in figure 4.11(c), we observe that, for conductive nanoparticles, there is a specific frequency at which the scattering efficiency is maximized. The attentive reader will notice that this peak occurs precisely at the SPR frequency, ω_{LSP}.

In fact, the LSP resonance is responsible for the peaks observed in the extinction, absorption, and scattering spectra of AgNPs. These peaks indicate the interaction between the incident light and the free electrons on the surface of the nanoparticles, resulting in a resonance that is visible in the presented graphs.

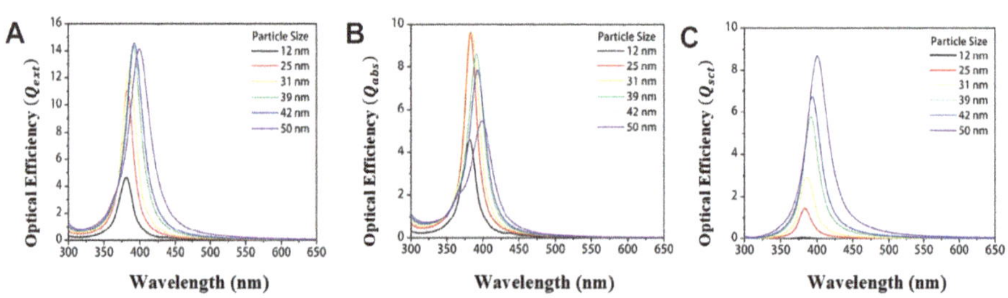

Figure 4.11. (A–C) Calculated UV–VIS spectra for AgNPs displaying controllable sizes from 12 to 50 nm as depicted in the insets: (A) optical efficiencies for extinction (Q_{ext}), (B) absorption (Q_{abs}), and (C) scattering (Q_{abs}). The optical efficiency is the ratio between the cross-section and the geometric area of the particle. [18] John Wiley & Sons. Copyright 2018 Wiley-VCH Verlag GmbH & Co. KGaA, Weinheim.

An important point in these spectra is the *redshift* in ω_{LSP} as the particle size increases. The Lorentz–Lorenz relation for polarizability leads to the Fröhlich condition, which does not predict the dependence of the LSP resonance frequency on the nanoparticle size (equation (4.46)). However, when we describe the polarizability including the radiation correction term, the resonance condition becomes a function of the particle size. The new resonance condition becomes

$$\mathrm{Re}\,\{\varepsilon_{\mathrm{c}} + 2\varepsilon_{\mathrm{m}}\} = \mathrm{Re}\left\{(\varepsilon_{\mathrm{c}} - \varepsilon_{\mathrm{m}})\left[(ka)^2 + \frac{2i}{3}(ka)^3\right]\right\}. \tag{4.59}$$

The resonance frequency can be graphically obtained as the solution to the above equation, that is, the value that satisfies the equality.

Thus, we can understand the redshift[3] as a radiative damping caused by the self-interaction of the dipole, which reduces the natural oscillation frequency of the dipole. Another factor contributing to the redshift is the weakening of the internal field with the increase in the particle's radius. Since $\mathbf{E}_{\mathrm{in}}$ has a Coulombian origin, it decreases with the increase in the distance between the charges, and consequently, the natural oscillation frequency of the charges decreases, leading to the redshift.

Using α_{rad} to calculate the scattering and extinction cross-sections, we can obtain results very close to those in figure 4.11 using the dipolar approach. Figure 4.12 shows the extinction cross-sections (4.12(a)) and scattering cross-section (4.12(b)) for AgNPs of three different sizes. The cross-sections were constructed from the optical properties of silver obtained by Johnson and Christy [17].

Figure 4.12(a) illustrates the extinction cross-section, while figure 4.12(b) illustrates the scattering cross-section. Note that for each radius, there is a resonance frequency that occurs at longer wavelengths (indicated in the figure) with the increase in the radius. These results are very close to those shown in figure 4.11, obtained from the discrete dipole approximation (DDA) [18].

A very different scenario occurs in figure 4.12(c), which shows the extinction cross-section calculated using the Lorentz–Lorenz relation. In addition to not observing the redshift with the increase in the particle radius, all the peaks occur at $\lambda_{\mathrm{LSP}} = 384$ nm, a value that satisfies the Fröhlich condition, and the cross-section values are significantly higher. This discrepancy in the cross-section values occurs because the self-interaction includes an imaginary term that calculates the energy dissipated radiatively before contributing to coherent scattering.

Beyond the size, the shape of the nanoparticles also significantly influences their extinction, scattering, and absorption spectra. A study conducted by Mock *et al* [18] experimentally demonstrated this relationship, in addition to providing an opportunity to understand the concept of cross-section and its relationship with the geometric section of the nanoparticles.

[3] The redshift is also influenced by the effects of retardation. For larger particles, the interaction with the incident electromagnetic field can no longer be treated solely in the quasi-static regime, and the resonance begins to depend on the distribution of the oscillating current over the volume of the nanoparticle. As a result, the plasmonic resonance shifts to lower frequencies, and scattering becomes more dominant than absorption, as evidenced in figure 4.11.

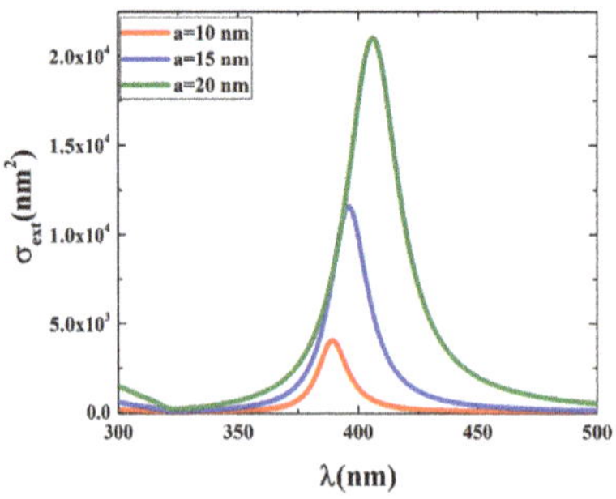

(a) Extinction cross section by α_{rad}.

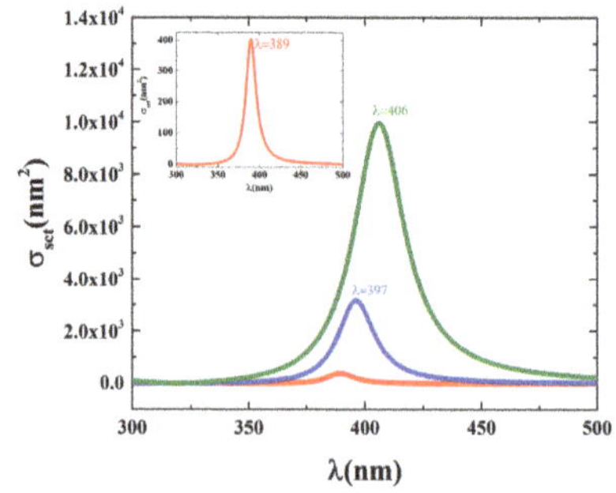

(b) Scattering cross section by α_{rad}.

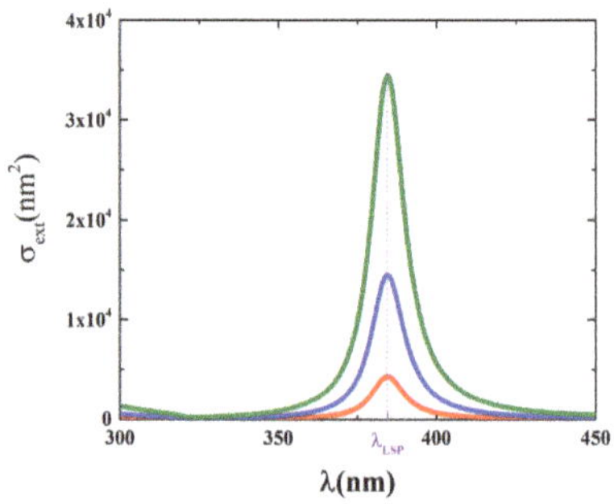

(c) Extinction cross section by α_{CM}.

Figure 4.12. Cross-section graphs for AgNPs as a function of wavelength, calculated using the dipolar model with the radiation correction for polarizability ((a) and (b)) and without the radiation correction for polarizability (c). (a) Extinction cross-section. (b) Scattering cross-section. (c) Extinction cross-section obtained using the Lorentz–Lorenz (Clausius–Mossotti) relation. Particle size: Solid red line: $a = 10$ nm, solid blue line: $a = 15$ nm, solid green line: $a = 20$ nm. $\lambda_{LSP} = 385$ nm. Optical constants obtained from reference [17].

In the study, the optical spectra of AgNPs with different shapes were compared using two techniques: dark-field microscopy (DFM) and transmission electron microscopy (TEM).

DFM illuminates the sample with white light at an oblique angle, while the collected light excludes direct illumination (figure 4.13(a)). This allows visualization of only the light scattered elastically by the nanoparticles, creating a dark background. Figure 4.13(b) shows a DFM image, where AgNPs appear as colored dots due to surface plasmon resonance. Figure 4.13(c), obtained by TEM, reveals the actual shapes and sizes of these nanoparticles. Since both images are at the same scale, the comparison clearly illustrates the difference between the optical cross-section and the geometric cross-section.

In DFM, the scattering intensity depends on the scattering cross-section, which represents the effective area of interaction of the nanoparticle with the incident light. This cross-section can be significantly larger than the geometric cross-section due to localized plasmon resonance, which amplifies the optical response at specific

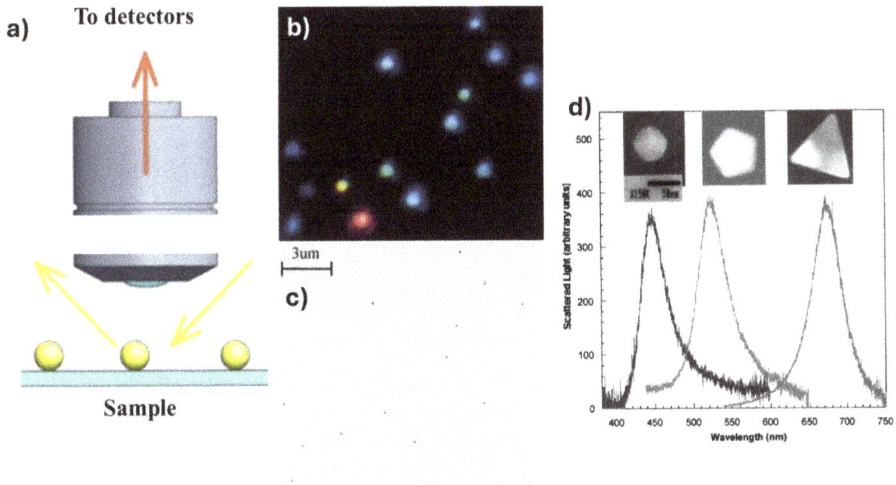

Figure 4.13. (a) Schematic illustrating the principle of DFM. White light is incident on the sample at a very high angle so that it is not directly collected by the microscope objective used to gather light for the detectors; only the light scattered by the sample is collected. [20] John Wiley & Sons. Copyright 2012 WILEY-VCH Verlag GmbH & Co. KGaA, Weinheim. (b) Colored DFM of AgNPs. (c) Bright-field TEM image of the same sample region. The dark-field image reveals the scattering cross-section of the AgNPs, whereas the TEM image shows their geometric area. (d) Dark-field spectrum (bottom) and TEM image (top) of AgNPs with different shapes. (b)–(d) Reprinted from [18] with the permission of AIP Publishing.

wavelengths. TEM, on the other hand, reveals only the actual physical size of the nanoparticles.

Figure 4.13(d) illustrates the relationship between the geometric shape of AgNPs and their spectral responses. The graph displays three distinct optical spectra, each corresponding to an individual nanoparticle observed in DFM. The TEM images above the spectra show that triangular nanoparticles exhibit resonance in the red region, pentagonal ones in the green, and spherical ones in the blue.

This result demonstrates that the shape of the nanoparticles directly influences their optical properties, as plasmon resonance depends on the distribution of surface charges and the collective oscillation modes of the conduction electrons. Moreover, anisotropic nanoparticles can exhibit multiple resonances due to the excitation of different plasmonic modes.

Although we can qualitatively interpret this resonance dependence on particle shape using the concepts discussed in this chapter, a rigorous analysis requires more sophisticated numerical calculations. For a deeper exploration of this topic, we recommend reading reference [19].

Within the OT theory, the increase in the extinction cross-section plays a key role, as it is closely related to the extinction forces, which characterize the non-conservative forces arising from the interaction between conductive nanoparticles and light. As we will see in the next section, the charge distribution plays a crucial role in creating stable trapping traps for conductive nanoparticles. The induced surface charges are related to the gradient force, which plays an essential role in the

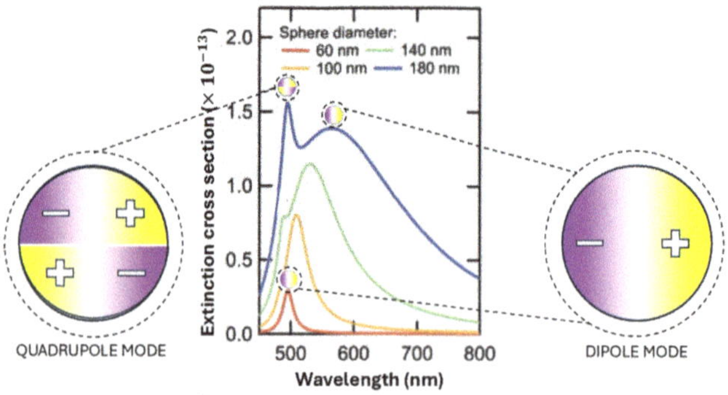

Figure 4.14. Extinction cross-section for AgNPs of different sizes calculated using Mie theory. For nanoparticles with diameters of 60 nm (solid red line) and 100 nm (solid orange line), there is a single peak corresponding to the dipolar mode. As the diameter increases to 140 nm (solid green line) and 180 nm (solid blue line), the peaks shift to the right (redshift of the dipolar mode), and a second peak appears on the left, characterizing the quadrupole mode. Highlighted is a schematic of the induced charge distribution for each mode. [20] John Wiley & Sons. Copyright 2012 WILEY-VCH Verlag GmbH & Co. KGaA, Weinheim.

capture and manipulation of nanoparticles. This force, which is conservative and attractive (under conditions that we will detail in the next section), arises from the interaction between the charges induced on the surface of the nanoparticle and the electric field of the incident light.

The distribution of these charges on the nanoparticle surface strongly depends on its size relative to the wavelength of light. For small particles (compared to λ), the optical response is well described by the dipole approximation, in which the charges organize into two opposite regions, as illustrated in the right part of figure 4.14. However, as the nanoparticle size increases, higher-order modes are activated, altering the charge distribution and, consequently, how the particle interacts with the optical field.

The central part of figure 4.14 presents the extinction cross-section for spherical AgNPs of different sizes. For smaller particles, a single well-defined peak is observed, corresponding to the dipolar mode. For larger particles (e.g. 180 nm), a second peak appears at shorter wavelengths, indicating the excitation of the quadrupolar mode.

This effect is directly reflected in the distribution of surface charges. As shown in the left part of figure 4.14, in the quadrupolar mode ($\lambda \approx 480$ nm), the charges are no longer organized as a simple dipole (as occurs for $\lambda \approx 580$ nm), but are instead distributed in four alternating regions of positive and negative charge. This implies that the gradient force on the nanoparticle must take into account the interaction between the optical field and this new charge configuration.

Thus, as the wavelength becomes shorter relative to the particle size, the dipole approximation becomes inadequate, and higher-order modes, such as the quadrupolar and others, must be considered. This transition directly affects how the gradient force acts on the particle, influencing the stability and efficiency of optical traps.

4.5 Radiation forces and optical trapping of plasmonic particles

When considering metallic particles, our initial intuition suggests that optical trapping of such materials is unlikely. The very existence of the gradient force in conducting particles appears counterintuitive at first since refraction, often regarded as the fundamental mechanism behind optical trapping[4], is absent. Indeed, conducting particles do not support the propagation of electromagnetic waves within their volume; instead, incident light is predominantly reflected at the particle's surface, leading to strong radiation pressure. While this pressure exerts a force on the particle, it does not generate a restoring optical potential as the gradient force does in dielectric particles. Within the framework of geometrical optics, this suggests that only repulsive forces should arise when light interacts with conducting particles.

A keen reader might question whether we could consider conducting particles as a limiting case of an absorbing particle, in which the absorption coefficient tends to infinity, thus allowing the use of the ideas discussed in chapter 3, section 3.3.2, considering that refraction corresponds to the penetration depth in the material. However, this approach does not apply. In section 3.3.2 it is assumed that the medium supports the propagation of electromagnetic waves and that, as the wave propagates, it is absorbed by the material. In conductors, however, the dispersion relation prevents wave propagation within the material, making it impossible to use this model, even as a limiting case.

The inapplicability of the model becomes evident when we look at typical metals such as gold or silver, which have a refractive index (real part) smaller than 1 in almost the entire visible and IR spectrum. If we directly applied the absorption model, we would erroneously conclude that only repulsive forces are induced on the particle. However, it is experimentally known that attractive forces can arise in metallic particles, such as gold, as reported in [21–24]. This contradiction suggests that a mechanism distinct from refraction must be acting in the interaction between light and metallic particles, allowing the emergence of attractive forces in these particles.

If geometrical optics does not allow us to understand how an attractive force can arise in a conducting particle, let us take the opposite approach and resort to the dipolar approximation to gain insight. Since the geometrical optics framework predicts only repulsive forces, we turn to the dipolar approximation, which better describes how conducting nanoparticles interact with light at these scales. Historically, the first optical trapping of a conducting particle was achieved precisely in this optical regime, in 1994, by Svoboda and Block [23]. Using gold nanoparticles (AuNPs) with a radius of 36 nm immersed in water, the experiment was conducted in a conventional OT setup with $\lambda = 1047$ nm. The authors observed that the optical trap formed for the AuNPs was about seven times stronger than that for latex particles of the same size under the same experimental conditions. This increase in trapping efficiency was accompanied by a similar increase in the polarizability of

[4] In dielectric particles, the refraction of light rays redirects photon momentum, creating a recoil force that drives the particle toward the region of higher photon flux near the laser focus.

AuNPs compared to latex particles. According to the authors, trapping occurs due to the similarity between the systems: in the Rayleigh regime ($a \ll \lambda$), both types of particles behave as induced dipoles, with AuNPs exhibiting a stronger optical trap due to their higher polarizability.

Indeed, when analyzing the interaction of conducting nanoparticles with light in section 4.4, we observed that their behavior is analogous to that of dielectric particles, in the sense that both become polarized, and the polarization charges interact with the electromagnetic field. From a practical standpoint, there is no fundamental difference between polarization charges arising from the displacement of the electronic cloud relative to the positive nucleus (as in dielectrics) and those resulting from the displacement of free electrons in conductors. As long as $a \ll \lambda$, the particle essentially behaves as an induced dipole and interacts with light accordingly. However, there is a fundamental difference: while dielectric particles exhibit a dipole forced by the incident light, metallic particles display a resonant behavior due to the response time of free electrons. This resonance amplifies the induced charges and, consequently, the polarizability of metallic particles, justifying the results observed by Svoboda and Block.

Beyond the simple polarization of the particle, light can couple to free electrons on the surface of the conductor, exciting LSPs. These plasmons introduce new forms of interaction between light and the particle, significantly modifying optical forces. In section 4.4, we saw that the LSP drastically alters the behavior of the scattering cross-sections of conducting particles (figure 4.12) compared to the behavior observed for dielectric particles (figure 2.8(c)). This difference arises from the way charges are induced in metallic particles. In particular, the LSP modifies the polarizable response of conducting nanoparticles with respect to the frequency of light (figure 4.9) and also as a function of particle size, since the resonance frequency ω_{LSP} depends on the nanoparticle radius (equation (4.59)).

In this section, we will apply the ideas discussed in section 4.4 in conjunction with the concepts from section 2.2 to understand how optical forces arise and act on conducting particles. In particular, we will investigate the role of LSP resonance in modifying optical forces and its implications for the optical trapping of metallic nanoparticles. Although the use of the dipolar approximation imposes limitations on the size range in which the model accurately describes the physical results, its understanding allows us to qualitatively grasp the processes involved in the creation of optical traps for metallic particles in other optical regimes. This is because the attractive force induced in a metallic particle, in the context of conventional OT, will always have a Coulombian origin, regardless of the particle size [24].

4.5.1 Polarizability

Throughout chapter 2, we discussed the emergence of optical forces when a dielectric particle interacts with light. In the interaction between light and particles in the Rayleigh regime, two main optical forces arise: the extinction force, which results from the transfer of momentum from the electromagnetic wave to the particle, and the gradient force, which emerges from the electrostatic interaction between the light electric field and the charges induced in the particle. The extinction force is associated with the imaginary part of the polarizability, whereas the gradient force is related to its real part.

Svoboda's work demonstrated that, in the quasi-static approximation ($a \ll \lambda$), the optical forces acting on conducting particles exhibit behavior similar to that observed in dielectric particles since, in both cases, the induced surface charges interact with the incident electric field. In particular, these two types of forces—the extinction force (comprising the scattering and absorption forces) and the gradient force—are present in both dielectric and conducting particles.

However, as discussed in section 4.4, the polarization mechanism in conducting particles differs from that in dielectric particles, leading to significantly different optical effects. For instance, as previously mentioned, the cross-sections exhibit a preferential frequency at which they reach a maximum (figure 4.12), unlike dielectric particles, where the cross-sections grow monotonically with ω (figure 2.8). This occurs because the dielectric response of dielectric particles does not present sharp resonances, whereas conducting particles exhibit plasmonic resonances. This change in the behavior of the cross-sections indicates that the extinction forces will behave differently in conducting particles compared to dielectric ones. At the LSP resonance, conducting particles experience a significantly stronger extinction force than at other frequencies, enhancing light scattering and absorption. This effect can hinder the stability of optical traps, as the gradient force may not be sufficient to overcome the radiation force generated by extinction.

To understand the role of LSP resonance in the gradient force, we must first discuss some aspects of the particle polarizability, many of which have already been addressed in section 4.4. Here, we contextualize these aspects to better understand optical forces. While the polarizability of dielectric particles can be described using both the Lorentz–Lorenz (Clausius–Mossotti) equation and the radiation correction, in the case of conducting particles, a complete description requires the inclusion of the radiation correction. This is because conducting particles undergo LSP resonance, which influences the behavior of polarizability across the entire visible spectrum and the near-infrared.

Of course, for $ka \ll 1$, the polarizability descriptions given by the Lorentz–Lorenz equation and the radiation correction provide similar values, making the Lorentz–Lorenz equation a good approximation, especially for small particles and wavelengths in the infrared spectrum, as in Svoboda's work. However, particularly for wavelengths in the visible spectrum, the use of the correction is necessary. Throughout this section, we will use the polarizability with the radiation correction, including the $(ka)^2$ term in the denominator, which plays a fundamental role in the accurate description of ω_{LSP}, as indicated in equation (4.59). Thus, the polarizability of a conducting particle is given by

$$\alpha_{\mathrm{d}} = \frac{\alpha_{\mathrm{CM}}}{1 - \dfrac{\varepsilon_{\mathrm{c}} - \varepsilon_{\mathrm{m}}}{\varepsilon_{\mathrm{c}} + 2\varepsilon_{\mathrm{m}}}\left[(ka)^2 + \dfrac{2i}{3}(ka)^3\right]}, \tag{4.60}$$

where α_{CM} is given by equation (4.42), and, as always, k represents the wavenumber in the medium.

Equation (4.60) is more effective in describing the polarizability of conducting particles than α_{CM}, as it accounts for the variation of ω_{LSP} with the particle radius. As discussed in section 4.4, several factors cause a redshift in the resonance frequency as the particle radius increases, one of which is self-interaction—that is, the interaction of the particle with its own scattered field. Equation (4.60) accurately describes the polarizability within a specific size range, whose upper limit depends on the optical properties of both the metal and the surrounding medium. For example, for gold particles in water, this limit is around 50 nm. For $a > 50$ nm, the equation tends to overestimate the resonance redshift, leading to an inaccurate description of the polarizability.

As discussed in section 2.2, for optical trapping to occur, it is necessary that $\mathrm{Re}\{\alpha_d\} > 0$. In the case of dielectric particles, this condition is satisfied when $\mathrm{Re}\{\varepsilon_p\} > \varepsilon_m$, or equivalently, when $n_p > n_m$, where n_p is the real part of the particle refractive index. This implies that the surface charges induced on the particle are distributed in such a way that the electric field at the position of the positive charges is stronger than at the position of the negative charges, resulting in a force that directs the particle toward regions of higher field intensity.

For conducting particles, the coupling between free electrons on the particle's surface and the incident electric field introduces a resonance condition that influences how charges are induced. In particular, there is a relationship between the frequency of the incident field, ω, and the particle's LSP resonance frequency, ω_{LSP}. This relationship plays a role analogous to that of the relative permittivity in dielectrics and can be visualized by comparing figures 2.3 and 4.9.

Note that many metals have a refractive index with a real part smaller than 1. If the trapping condition were analogous to that of dielectric particles, no conducting particle could be optically trapped. For a metal described by the Drude model, optical trapping occurs when $\omega < \omega_{LSP}$. However, this is not a general condition, as some metals are not well described by the Drude model across the entire visible and ultraviolet spectrum. In certain cases, $\mathrm{Re}\{\alpha_d\} > 0$ can still be observed even for $\omega > \omega_{LSP}$ due to a more complex dependence between ε_c and ω than that predicted by the Drude model. A notable example is gold, which, due to interband transitions and other effects, exhibits $\mathrm{Re}\{\alpha_d\} > 0$, even for $\omega > \omega_{LSP}$.

However, ω_{LSP} is not a fixed parameter for each metal, as it depends on the particle radius. Thus, for each particle size, the polarizability α_d exhibits a distinct behavior, with a specific wavelength λ at which $\mathrm{Im}\{\alpha_d\}$ reaches its maximum value, characterizing the LSP resonance. This behavior is illustrated in figure 4.15, which shows the calculated polarizability for AgNPs with three different radii. The values of α_d were obtained using equation (4.42), with optical parameters extracted from reference [17].

In the case of silver, it is observed that the wavelength for which $\mathrm{Re}\{\alpha_d\} = 0$ coincides with the one where $\mathrm{Im}\{\alpha_d\}$ reaches its maximum value, as illustrated in figure 4.15. However, this coincidence is not a general rule for all metals. In many cases, the frequency at which $\mathrm{Re}\{\alpha_d\}$ changes sign is slightly shifted relative to the

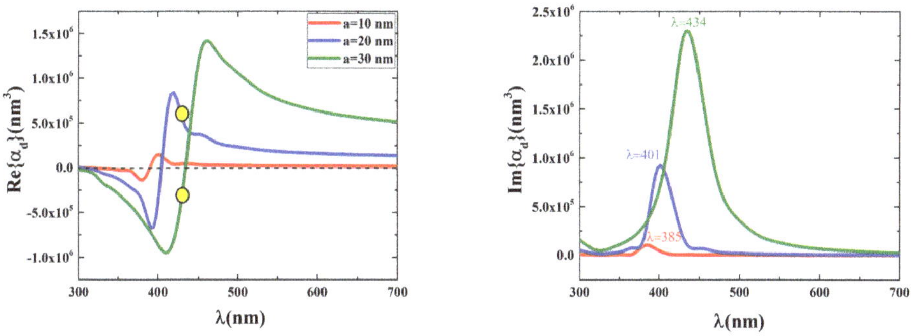

(a) Real part of the polarizability. (b) Imaginary part of the polarizability.

Figure 4.15. Polarizability of a silver particle in water as a function of the beam wavelength for different particle sizes. $a = 10$ nm (red solid line), $a = 20$ nm (blue solid line), $a = 30$ nm (green solid line). (a) Real part of the polarizability. The yellow points indicate $\lambda = 430$ nm. (b) Imaginary part of the polarizability. The resonance wavelengths are indicated in the figure. Parameters: $P = 100$ mW, $w_0 = 2$ μm, $\varepsilon_m = 1.77$; optical data for silver obtained from reference [17].

LSP resonance frequency due to the presence of optical losses in the material. Moreover, depending on the metal's electronic structure and the influence of additional mechanisms such as interband absorption, multiple wavelengths may exist for which $\text{Re}\{\alpha_d\} = 0$. This indicates that the vanishing of the real part of the polarizability is not, by itself, an absolute criterion for identifying the LSP resonance.

This analysis highlights how the LSP resonance directly influences the polarizability and, consequently, the optical forces exerted on conducting particles. For example, for a beam with $\lambda = 430$ nm, the polarizability of a silver particle with a radius of $a = 20$ nm is positive, whereas for $a = 30$ nm, the polarizability becomes negative (yellow circle in figure 4.15(a)). This behavior contrasts with that of dielectric particles, where an increase in radius results in a greater magnitude of $\text{Re}\{\alpha_d\}$ but does not cause a sign inversion.

4.5.2 Gradient forces in conducting particles

Since the LSP resonance alters the particle's polarizability behavior, it is natural to expect that the gradient force will also be affected. Before analyzing this influence, it is important to understand how the gradient force arises and acts specifically on conducting particles.

From a fundamental perspective, the mechanism of gradient force generation in metals does not differ significantly from that observed in dielectric particles. The incident electric field induces a polarization in the particle, leading to charge separation: positive charges accumulate at one pole, while negative charges accumulate at the opposite pole. As discussed in section 4.4, this process results from the migration of free electrons in response to the applied field.

As established in section 2.2.3, the gradient force arises when there is a spatial gradient in the intensity of the electromagnetic field. Consequently, the electric field at the position of the positive charges, $\mathbf{E}_i(\mathbf{r}_+)$, differs from that at the position of the negative charges, $\mathbf{E}_i(\mathbf{r}_-)$. Defining $\mathbf{F}_+(\mathbf{r}_+) = q\mathbf{E}_i(\mathbf{r}_+)$ as the force exerted on the positive charges and $\mathbf{F}_-(\mathbf{r}_-) = -q\mathbf{E}_i(\mathbf{r}_-)$ as the force on the negative charges, the resulting gradient force on the particle can be written as

$$\mathbf{F}_G(\mathbf{r}_\mathrm{d}) = \mathbf{F}_+(\mathbf{r}_+) + \mathbf{F}_-(\mathbf{r}_-), \tag{4.61}$$

where $\mathbf{r}_\mathrm{d}$ represents the centroid position of the particle.

Using equations (2.80) and (2.82), we can then express the gradient force on a conducting particle as

$$\mathbf{F}_G(\mathbf{r}_\mathrm{d}) = \frac{1}{4}\varepsilon_0\varepsilon_\mathrm{m}\mathrm{Re}\{\alpha_\mathrm{d}\}\nabla|\mathbf{E}_i(\mathbf{r}_\mathrm{d})|^2. \tag{4.62}$$

This equation has an analogous structure to the gradient force on dielectric particles (equation (2.83)). The force will be attractive, directing the particle toward regions of higher field intensity, when $|\mathbf{F}_+| > |\mathbf{F}_-|$, or equivalently, when $\mathrm{Re}\{\alpha_\mathrm{d}\} > 0$. On the other hand, when $|\mathbf{F}_-| > |\mathbf{F}_+|$, i.e. $\mathrm{Re}\{\alpha_\mathrm{d}\} < 0$, the gradient force becomes repulsive, pushing the particle away from regions of higher field intensity.

The charge distribution on the conducting particle is directly influenced by the coupling of free electrons with the electromagnetic field. For fields with frequencies below the resonance frequency, the induced charge distribution generates an attractive gradient force. Conversely, when the field frequency exceeds the resonance frequency, the charge distribution reverses, resulting in a repulsive gradient force. This behavior reflects, as expected, the sign of the particle polarizability.

Figure 4.16 illustrates this effect, showing the forces acting on two silver particles, one with a radius of $a = 20$ nm (figure 4.16(a)) and another with $a = 30$ nm (figure 4.16 (b)), both subjected to a Gaussian beam with a wavelength of $\lambda = 430$ nm. The increase in particle size shifts the resonance frequency ω_LSP, causing the smaller particle to experience an attractive gradient force at $\lambda = 430$ nm, while the larger particle undergoes a repulsive force.

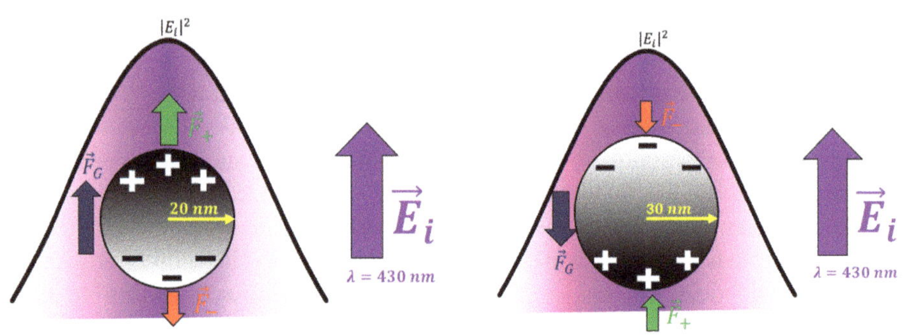

(a) Coulomb forces on Ag particles with $a = 20$ nm.

(b) Coulomb forces on Ag particles with $a = 30$ nm.

Figure 4.16. Distribution of Coulomb forces generated by the induced charges on the silv particle in $\lambda = 430$ nm. (a) $a = 20$ nm. (b) $a = 30$ nm.

A fundamental point to highlight is that, in figure 4.16, both particles are made of the same material, have the same optical properties, are immersed in the same medium (water), and interact with the same light beam. The only variable between the two scenarios is the particle radius, which alone is sufficient to reverse the nature of the gradient force. This subtlety demonstrates that, although optical properties are essential, the resulting force is determined by the coupling between the particle's free electrons and the light electric field. In other words, it is the excitation of surface plasmons that dominates the interaction. For this reason, conducting particles are often referred to as *plasmonic particles*.

This behavior highlights a unique characteristic of plasmonic particles compared to dielectrics. While in dielectric particles, the sign of the gradient force is primarily governed by the relative permittivity and remains largely independent of particle size in the dipole approximation, in metallic particles, increasing the radius directly alters the LSP resonance condition, modifying the polarizability and consequently reversing the force direction. This effect is a direct manifestation of the coupling between free electrons and the incident field, reinforcing the resonant nature of the interaction. Therefore, controlling the particle size becomes a critical parameter in OT experiments with metallic nanoparticles, allowing tuning of the gradient force from attractive to repulsive simply by adjusting the radius, without the need to change the material or the incident beam.

The role of plasmonic resonance becomes even more evident when analyzing the optical potential induced during the interaction of light with plasmonic particles. Although the definition of optical potential is similar to that adopted for dielectric particles, its interpretation in metallic particles needs to be expanded due to the presence of LSP resonance.

In dielectrics, the optical potential is understood as the effective potential of the electrostatic interaction between the light's electric field and the induced polarization charges, resulting in a dipolar potential that governs optical trapping. However, in plasmonic particles, beyond this classical interaction, it is necessary to explicitly consider the contribution of LSP resonance. This is because the free electrons on the particle surface can oscillate collectively in response to the incident electric field, modifying the particle response in a resonant manner.

Thus, the optical potential in plasmonic particles can be defined as: *the effective potential associated with the interaction between the incident electric field and the induced polarization charges, including the resonant modification of the particle's response due to the excitation of localized surface plasmons*. Mathematically, this potential is expressed as

$$U_{\mathrm{op}}(\mathbf{r}_{\mathrm{d}}) = -\frac{1}{4}\varepsilon_0\varepsilon_{\mathrm{m}}\,\mathrm{Re}\{\alpha_{\mathrm{d}}\}|\mathbf{E}_i(\mathbf{r}_{\mathrm{d}})|^2 = -\frac{1}{2}\frac{n_{\mathrm{m}}}{c}\,\mathrm{Re}\{\alpha_{\mathrm{d}}\}I(\mathbf{r}_{\mathrm{d}}), \tag{4.63}$$

where the last equality results from the use of equation (2.84).

This expression is analogous to the dielectric case (equation (2.88)). However, for plasmonic particles, the polarizability α_{d} exhibits a strong frequency dependence, being significantly modified by the coupling between the incident electric field and the free electrons of the particle. As a consequence, the behavior of the optical potential directly depends on the relationship between the field frequency and the LSP resonance frequency (ω_{LSP}). Considering the dielectric response described by the Drude model, this behavior can be summarized as follows:

- For $\omega < \omega_{\mathrm{LSP}}$, the polarizability exhibits $\mathrm{Re}\{\alpha_{\mathrm{d}}\} > 0$, resulting in an attractive optical potential. In this regime, the particle is drawn toward regions of higher field intensity, allowing the formation of stable optical traps.
- For $\omega > \omega_{\mathrm{LSP}}$, the polarizability becomes negative, with $\mathrm{Re}\{\alpha_{\mathrm{d}}\} < 0$. Under this condition, the optical potential behaves as a repulsive barrier, pushing the particle away from high-intensity regions and forcing it to migrate toward lower-intensity field zones.

This behavior contrasts with that observed in dielectric particles, where the optical potential is always attractive, leading to particle confinement in regions of maximum field intensity (assuming $\varepsilon_{\mathrm{p}} > \varepsilon_{\mathrm{m}}$). In the plasmonic case, the stability of optical trapping critically depends on the light frequency relative to the particle's resonance, as well as being sensitive to the particle size and the surrounding dielectric medium.

In other words, while the optical potential in dielectric particles can be interpreted as a classical dipolar potential, in plasmonic particles, it acquires a resonant and highly dispersive character. This resonance, resulting from the coupling between the electric field and free electrons, allows the existence of both attractive and repulsive potentials, making the optical interaction much more dynamic and rich in phenomena dependent on experimental conditions.

As with polarizability, real metals may not strictly obey the criterion $\omega < \omega_{\mathrm{LSP}}$, to ensure a confining optical potential. A clear example is gold, whose optical response allows for an attractive potential even in the region $\omega > \omega_{\mathrm{LSP}}$. This reflects the fact that, within this frequency range, gold still exhibits $\mathrm{Re}\{\alpha_{\mathrm{d}}\} > 0$. Figure 4.17 illustrates the behavior of the optical potential for three real metals—silver (Ag), gold (Au), and copper (Cu)—whose dielectric constants were obtained from the experimental data in reference [17]. For each metal, we considered five different radii, and the potential was computed over the wavelength range from 400 to 1100 nm.

For comparison, we also included the idealized case in which the plasmonic particle is described by the Drude model, adopting $\omega_{\mathrm{p}} = 8 \times 10^{15}$ rad s^{-1} and $\gamma = 5 \times 10^{13}$ s^{-1}. The incident beam was modeled as a zero-order Gaussian with a beam waist of $w_0 = 2000$ nm and power $P = 100$ mW. The optical potential was calculated for the particle positioned at $(\tilde{x}, \tilde{y}, \tilde{z}) = (0, 0, 0)$, following the geometry and definitions from section 2.2.2.

Figures 4.17(a) to 4.17(c) show the optical potential for Ag, Au, and Cu particles, respectively, considering five different radii. The potential was obtained using equation (4.63). It is observed that, for each material, the optical potential is significantly modified by the particle size, highlighting the role of LSP resonance.

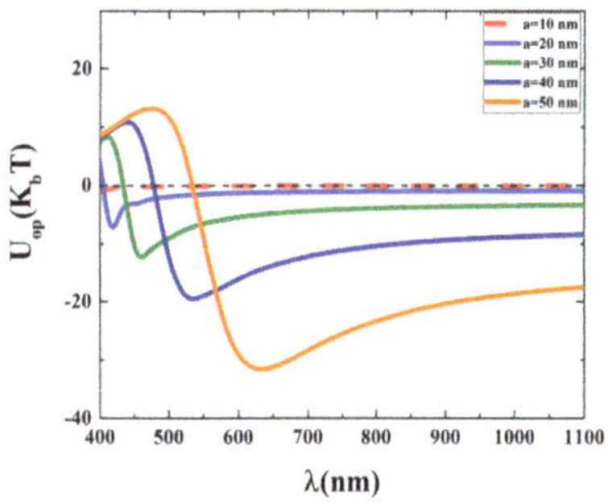

(a) Optical potential of silver particles.

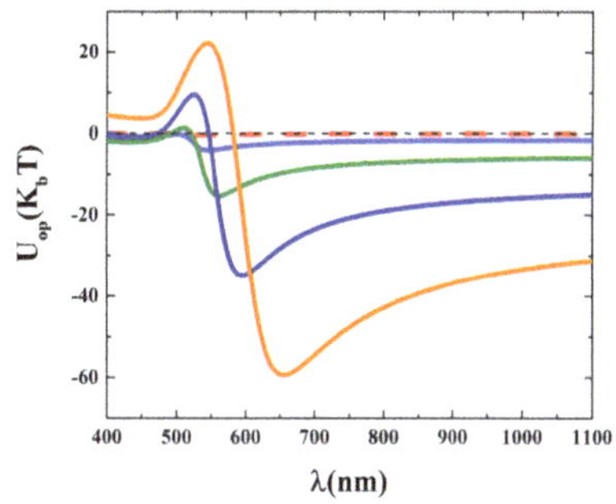

(b) Optical potential of gold particles.

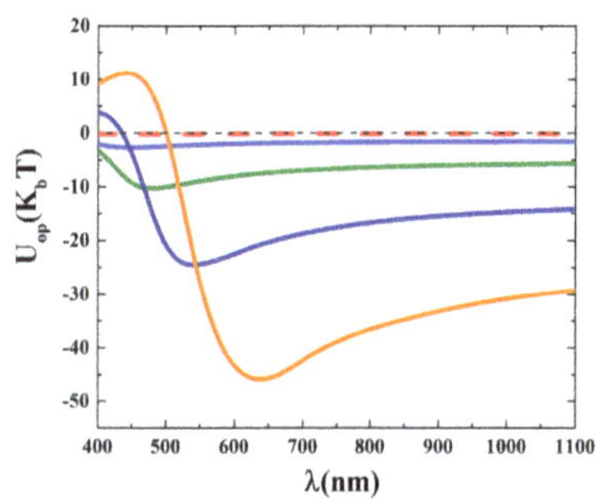

(c) Optical potential of copper particles.

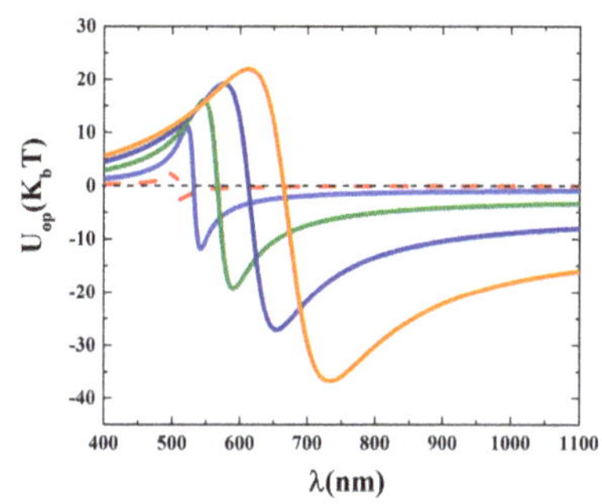

(d) Optical potential of Drude particles.

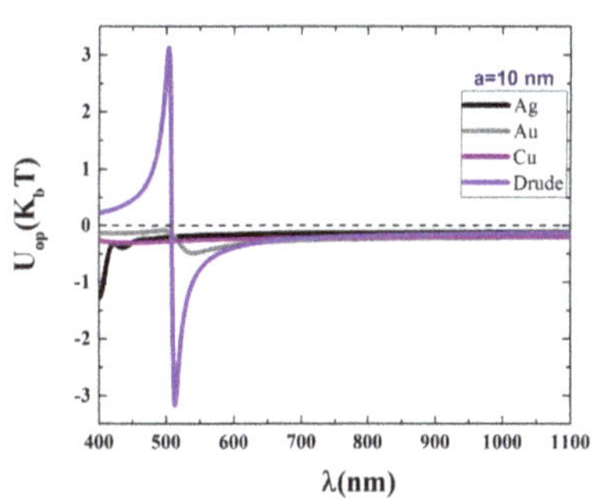

(e) Optical potentials for $a = 10$ nm.

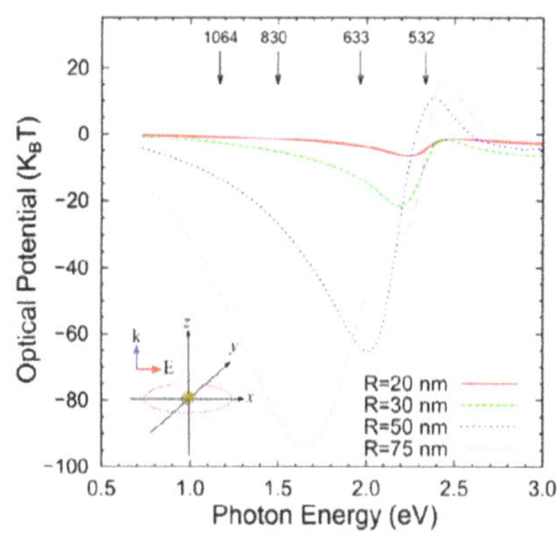

(f) Optical potential of gold particles through electrodynamics calculations.

Figure 4.17. Optical potentials induced by a zero-order Gaussian beam on plasmonic particles of different sizes calculated using the dipole approximation. (a) Ag. (b) Au. (c) Cu. (d) Drude particle. (e) Potential for the particles in (a)–(d) when $a = 10$ nm. (f) Optical potential for gold particles of different sizes calculated by solving Maxwell's equations within the Mie scattering framework, obtained with permission from [25], copyright 2015 American Chemical Society. **Parameters**: (a)–(e): $w_0 = 2.0\ \mu$m, $P = 100$ mW. $P = 10$ mW. Drude particle parameters: $\omega_p = 8 \times 10^{15}$ rad s^{-1}, $\gamma = 5 \times 10^{13}$ s^{-1}. Optical data for Ag, Au, and Cu obtained from reference [17] (f): NA = 0.7. $\mathbf{r}_d = (0, 0, 0)$.

In particular, we can extend the discussion from figure 4.16 to the optical potential by comparing silver particles with radii of 20 nm and 30 nm when $\lambda = 430$ nm. It is noted that while the potential for the 20 nm particle (solid blue line) is negative—characterizing a confining potential—the potential for the 30 nm particle (solid green line) becomes positive, assuming the character of a repulsive barrier.

However, for the other materials, Au and Cu, the potential remains negative at $\lambda = 430$ nm for particles with $a = 30$ nm, indicating the formation of a confining potential. This result highlights an important point: the nature of the optical potential is not defined solely by the particle size but also by the internal electronic structure reflected in the metal optical properties.

The influence of the electronic structure becomes even more evident when comparing the optical potential behavior of a real metal with that of a metal described by the Drude model, as illustrated in figure 4.17(e) for particles with $a = 10$ nm. Using the Fröhlich condition (equation (4.46)) as a criterion for resonance, we obtain the following resonant wavelengths for particles in water: $\lambda_{Ag} = 383$ nm, $\lambda_{Au} = 520$ nm, and $\lambda_{Cu} = 424$ nm. We also include the case of the Drude metal, for which $\lambda_{Drude} = 500$ nm.

If the optical properties of these metals strictly followed the Drude model over the entire analyzed range, no particle could be trapped at wavelengths shorter than its respective λ_{LSP}. Within the range of 400 to 1100 nm, and considering this purely Drude-like scenario, only silver should be optically trapped, regardless of the particle size.

However, the results show that Ag, Au, and Cu particles with $a = 10$ nm generate confining potentials throughout the entire analyzed λ range. This behavior contrasts with that of the Drude metal, where the potential is confining only for $\lambda > \lambda_{LSP}$.

This discrepancy arises because, when modeling the dielectric function using the Drude model, we consider only the contribution of free electrons, which are responsible for the LSP resonance. However, bound electrons also influence the dielectric response, especially in the ultraviolet-to-blue region, where photon energy is sufficient to induce interband transitions in metals like gold.

When the particle radius is large enough to shift the resonance to wavelengths where the effect of bound electrons can be neglected—that is, when the dielectric response is well described by the Drude model—the optical potential behavior starts resembling that of a Drude metal. This is the case for silver, whose bound electron effects are negligible within the analyzed range, for copper (when $a > 40$ nm), and for gold (when $a > 50$ nm).

In other words, for sufficiently large particles and optical traps operating in the visible or infrared, the expected behavior of the optical potential tends to follow the Drude regime: the potential will be confining for $\lambda > \lambda_{LSP}$ and act as a barrier for $\lambda < \lambda_{LSP}$. This result makes it clear that the condition $\lambda < \lambda_{LSP}$ implying a repulsive potential is not a general rule for all metals but rather a direct consequence of the Drude model, in which only free electrons are considered in the optical response.

When bound electron contributions are significant, as in real metals, this relationship can break down, allowing optical confinement even for $\lambda < \lambda_{LSP}$.

Unfortunately, there is a size range limitation in which we can describe the optical potential between plasmonic nanoparticles and light using the dipolar model. For dielectric particles, we have seen that the optical potential is well described by the dipolar model as long as the particle radius is smaller than the beam waist w_0. However, for plasmonic particles, this limitation does not apply in the same way because the coupling between free electrons and the electromagnetic field leads to the formation of a size-dependent resonance.

The uniformity of the electric field inside the nanoparticle suggests that the distribution of induced surface charges will be predominantly dipolar. However, the coupling between these charges and the incident field gives rise to a resonant dipole. Thus, our model must adequately describe this resonance. In principle, the resonance frequency does not depend on w_0 but only on the particle radius, the metal optical properties, and the dielectric medium in which the metal is immersed.

The resonance condition provided by equation (4.59) reasonably describes the variation of ω_{LSP} with particle size as long as the particle remains small. This is not surprising, given that equation (4.59) only considers the self-interaction effect of the dipole-generated field, essentially the near-field scattering. For larger particles, the scattered field is no longer well described by the dipolar approximation, and other effects, such as those discussed in the previous section, begin to influence the resonance.

Strictly speaking, our model adequately describes the optical potential for particles in the Rayleigh regime ($a < \lambda/20$). To validate our model within the validity range of the dipolar approximation, we compare our results with those obtained from a more robust numerical approach based on Mie theory and focused fields, as shown in figure 4.17(f).

Figure 4.17(f), reprinted from reference [25], presents electrodynamic calculations of optical forces acting on spherical gold particles with different radii ($a = 20$, 30, 50, and 75 nm) in water when illuminated by a focused laser beam. These calculations use the beam angular spectrum representation, considering only propagating components. Unlike the fields obtained in the paraxial approximation, the fields incident on the nanoparticle are exact solutions of Maxwell's equations. These fields are projected onto vector spherical harmonics and expanded in terms of electric and magnetic multipolar modes at the particle surface, with the projection performed numerically.

The force on the particle is then calculated within the framework of Mie theory, using the Wigner–Eckart theorem. The optical potential results from integrating this force, assuming that the particles are located at the center of the focal plane of a Gaussian beam focused by an objective with $\mathrm{NA} = 0.7$.

It is observed that our model (figure 4.17(b)) satisfactorily predicts the qualitative behavior of the optical potential for gold particles with radii up to $a = 50$ nm. Naturally, there are differences in potential depth, reflecting expected variations in local electric field intensities. However, our focus is on qualitative description—that is, in which frequency ranges the potential is confining or assumes the character of a barrier.

For particles with radii greater than 50 nm (not shown), our model can no longer accurately compute LSP resonance frequencies, resulting in discrepancies relative to the potential predicted by the exact solution.

All our discussion on the optical potential naturally extends to the gradient force. Using the geometry described in section 2.2.2, the components of the gradient force follow analogously from equations (2.85)–(2.87). As long as the field description using the zeroth-order Gaussian approximation remains valid, only the gradient force acts on the particle in the transverse direction to the beam propagation.

As in the case of dielectric particles, the transverse components of the gradient force, $\mathbf{F}_{Gx}$ and $\mathbf{F}_{Gy}$, are symmetric, reflecting the symmetry of the beam intensity. However, in addition to the magnitude, the direction of the gradient force also depends on the particle size, as a consequence of the LSP resonance.

Figure 4.18(a) illustrates the x-component of the gradient force as a function of the normalized displacement relative to the beam center for two spherical AgNPs in water, with different radii, $a = 20$ nm (blue dashed line) and $a = 30$ nm (green dashed line). The beam used is a zeroth-order Gaussian beam with $w_0 = 2\ \mu$m, $\lambda = 430$ nm, and power $P = 100$ mW.

This result highlights our discussion on the impact of particle size on polarizability and optical potential, directly influencing the gradient force. Although in both cases $\mathbf{F}_{Gx}$ is conservative—meaning the work done by the force is path-independent—only for the smaller particle ($a = 20$ nm) is the gradient force restorative, leading the nanoparticle to the optical axis ($x = 0$).

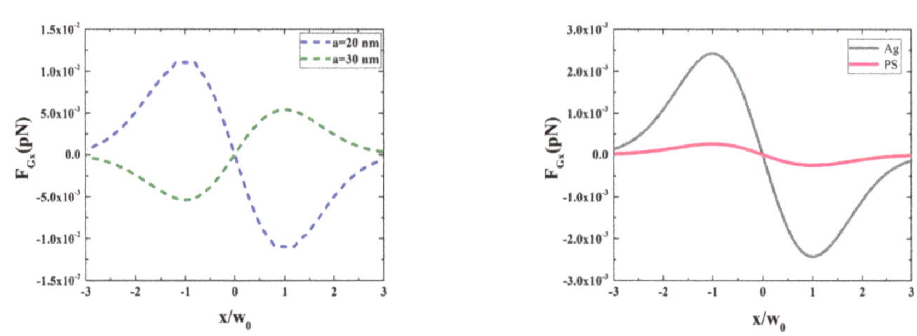

(a) Transversal component of the gradient force for Ag NPs.

(b) Comparison of the transversal component of the gradient force between Ag NPs and PS NPs.

Figure 4.18. (a) Variation of the x-component of the gradient force as a function of the normalized position for AgNPs of two different sizes: $a = 20$ nm (blue dashed line, where the gradient force is attractive) and $a = 30$ nm (green dashed line, where the gradient force is repulsive) at $\lambda = 430$ nm. (b) Comparison of the x-component of the gradient force for a plasmonic nanoparticle (silver, solid gray line) and a dielectric nanoparticle (polystyrene (PS NPs), solid pink line), both in water, with $a = 20$ nm and $\lambda = 1047$ nm. **Parameters:** $w_0 = 2\ \mu$m, $P = 100$ mW, $\varepsilon_m = 1.77$. Optical data for silver obtained from reference [17]. $\mathbf{r}_d = (x/w_0, 0, 0)$.

Besides making the gradient force repulsive under certain conditions, the LSP resonance also amplifies the force magnitude, enabling the creation of more stable optical traps for plasmonic particles.

Figure 4.18(b) illustrates this effect by comparing the x-component of the gradient force on a plasmonic AgNP and a dielectric PS NPs particle, both with the same radius ($a = 20$ nm), under a zeroth-order Gaussian beam with $\lambda = 1047$ nm and $P = 100$ mW.

It is observed that the gradient force on the AgNP is approximately 9.6 times larger than that on the polystyrene particle at the position $(x, y, z) = (\pm w_0/2, 0, 0)$, where the force magnitude is maximal. This result is of the same order of magnitude as that observed by Svoboda when comparing the gradient force between gold and latex particles, with an approximate difference of seven times, as discussed at the beginning of this section.

The most common method for quantifying the stability of optical trapping is through the trap stiffness or force constant, κ, defined as before

$$\kappa_{x_i} = -\frac{\partial F_{Gx_i}}{\partial x_i}\bigg|_{x_i = 0} \tag{4.64}$$

where $x_i = \{x, y, z\}$. Here, κ_{x_i} has the same interpretation as presented in section 2.2.3, meaning it quantifies the restoring force acting on the plasmonic particle in the optical trap, bringing it back to the equilibrium position after a small displacement in the x_i direction. For small displacements relative to the equilibrium position, the optical potential can be considered harmonic (figure 2.11). Figure 4.19(a) shows the behavior of the trap stiffness in the x direction as a function of experimentally controllable parameters.

In figure 4.19, we observe how κ_x varies with the wavelength λ for five different sizes of spherical AgNPs ($a = 10$ to 50 nm), reflecting the behavior of the optical potential shown in figure 4.17(a). As in the case of dielectric particles, negative values of κ can be interpreted as indicative of a barrier-type potential. Note that for each nanoparticle size, there is a specific wavelength that maximizes the trap stiffness. This result is complemented by figure 4.19(b), which shows the behavior of κ as a function of particle size for $\lambda = 430$ nm.

It is interesting to note that this behavior differs from what is observed for dielectric particles (figure 2.12), where $\kappa \propto a^3$. Here, we see that there is a nanoparticle size that maximizes the trap stiffness, as well as regimes in which $\kappa < 0$. These results are particularly useful for optimizing optical manipulation setups, as they allow for the identification of the most suitable parameters for achieving more efficient traps or designing traps within a specific force range of interest. This is particularly relevant for applications such as force spectroscopy on specific molecules, including DNA, RNA, proteins, and others [26–28].

Finally, figure 4.19(c) reinforces this application potential by showing that κ (and therefore the gradient force) remains linearly proportional to the laser power. This provides a useful degree of freedom for tuning the optical trap to suit the aforementioned applications.

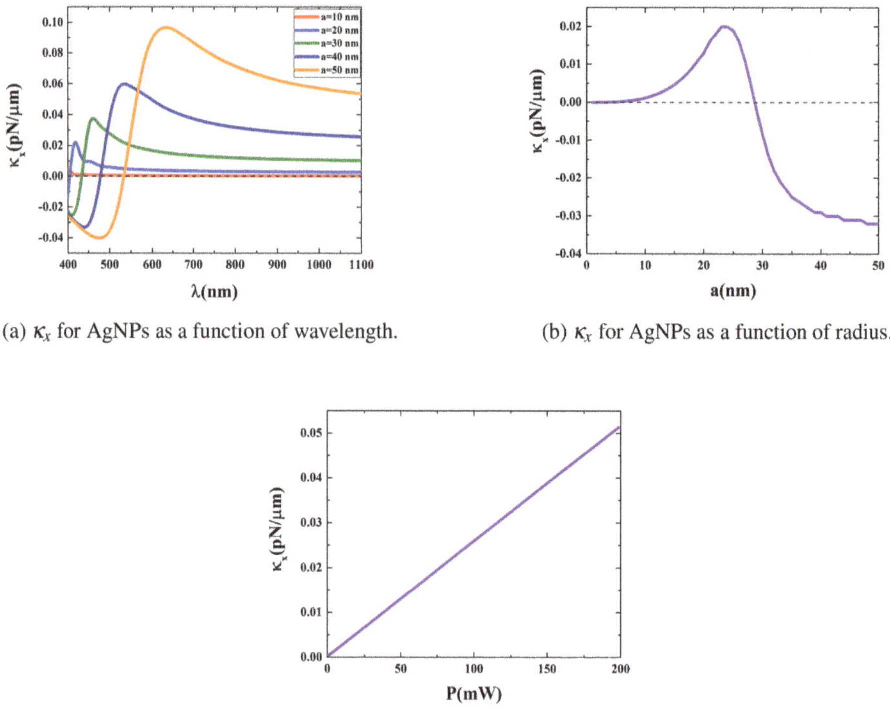

(a) κ_x for AgNPs as a function of wavelength.

(b) κ_x for AgNPs as a function of radius.

(c) κ_x for AgNPs as a function of power.

Figure 4.19. Trap stiffness in the x direction (κ_x) for AgNPs in water. (a) κ_x as a function of the wavelength for various sizes of AgNPs. $P = 100$ mW. (b) κ_x as a function of the AgNP radius. $P = 100$ mW, $\lambda = 430$ nm. (c) κ_x as a function of the beam power. $a = 25$ nm, $\lambda = 430$ nm. Parameters: $w_0 = 2\ \mu$m. Optical parameters for silver obtained from reference [17]. $\mathbf{r}_\text{d} = (0, 0, 0)$.

The origin of these behaviors of κ is, as the reader is already aware at this point, the LSP resonance. However, it is not only the gradient force that is modified by the LSP. As the more attentive reader may have noticed since we introduced the cross-sections for plasmonic particles, the extinction forces also exhibit distinct behaviors compared to those observed in dielectric particles due to the LSP.

Both the scattering cross-section and the absorption cross-section—and consequently, the extinction cross-section—are affected by the LSP resonance. As a result, the forces associated with radiation pressure also exhibit the signature of the LSP, making their behavior more complex in relation to particle size and wavelength compared to what we described for dielectric particles in section 2.2.4. Mathematically, the forces associated with radiation pressure can be defined by the following equations:

$$\mathbf{F}_\text{ext} = \frac{n_\text{m}}{c}\sigma_\text{ext}\langle \mathbf{S}_i(\mathbf{r}_\text{d}, t)\rangle = \frac{n_\text{m}}{c}\sigma_\text{ext} I(\mathbf{r}_\text{d})\hat{z}, \tag{4.65}$$

$$\mathbf{F}_\text{sct} = \frac{n_\text{m}}{c}\sigma_\text{sct}\langle \mathbf{S}_i(\mathbf{r}_\text{d}, t)\rangle = \frac{n_\text{m}}{c}\sigma_\text{sct} I(\mathbf{r}_\text{d})\hat{z}, \tag{4.66}$$

$$\mathbf{F}_{\mathrm{abs}} = \frac{n_{\mathrm{m}}}{c}\sigma_{\mathrm{abs}}\langle \mathbf{S}_i(\mathbf{r}_{\mathrm{d}}, t)\rangle = \frac{n_{\mathrm{m}}}{c}\sigma_{\mathrm{abs}} I(\mathbf{r}_{\mathrm{d}})\hat{z}, \tag{4.67}$$

where $\mathbf{F}_{\mathrm{ext}} = \mathbf{F}_{\mathrm{sct}} + \mathbf{F}_{\mathrm{abs}}$ and $I(\mathbf{r}_{\mathrm{d}})$ is the beam intensity at the centroid position of the particle, given by equation (2.78).

Physically, radiation pressure forces—or, as we referred to them in section 2.2.4, extinction forces—represent the rate of momentum removed from the electromagnetic field by the particle. These forces are non-conservative and, within the zeroth-order Gaussian approximation, have only a component in the longitudinal direction (along the z-axis).

For non-absorbing dielectric particles, we have seen that the extinction cross-section (and, consequently, the extinction force) is dominated by scattering, which allowed us to approximate $\mathbf{F}_{\mathrm{ext}} \approx \mathbf{F}_{\mathrm{sct}}$. However, for plasmonic particles, this is not always true: depending on the particle size and the wavelength of the incident beam, the absorption force may be significant or even exceed the scattering force. This reflects the fact that, in this configuration, the plasmonic particle absorbs the incident light more efficiently than it scatters it.

Figure 4.20(a) illustrates this effect by showing the magnitude of $\mathbf{F}_{\mathrm{sct}}$ (solid red line) and $\mathbf{F}_{\mathrm{abs}}$ (solid gray line) for two silver particles of different sizes ($a = 10$ nm and $a = 20$ nm). Note that for the smaller particle, $|\mathbf{F}_{\mathrm{sct}}| < |\mathbf{F}_{\mathrm{abs}}|$, whereas for the larger particle, $|\mathbf{F}_{\mathrm{sct}}| > |\mathbf{F}_{\mathrm{abs}}|$. In both cases, however, both forces contribute significantly to the total value of $\mathbf{F}_{\mathrm{ext}}$ (dashed purple line).

Although scattering and absorption forces have distinct physical interpretations, both act on the particle by accelerating it in the direction of beam propagation. For this reason, it is more convenient to work directly with the extinction force $\mathbf{F}_{\mathrm{ext}}$, which represents the resultant radiation pressure force in the longitudinal direction.

At resonance frequencies, the extinction force $\mathbf{F}_{\mathrm{ext}}$ is significantly amplified, directly reflecting the increase in the extinction cross-section. Figure 4.20(b) shows the behavior of $\mathbf{F}_{\mathrm{ext}}$ as a function of the radiation wavelength (measured in vacuum) for silver spherical particles in water with different sizes. It is noted that, at resonant wavelengths, the extinction force reaches values on the order of pN—an increase of 4–5 orders of magnitude compared to what is expected for a dielectric particle of the same size.

An intuitive way to understand this effect is to imagine the plasmonic nanoparticle as a sail being pushed by the light beam due to radiation pressure. The 'area' of this sail is given by the particle's extinction cross-section, which effectively represents how much the nanoparticle interacts with the electromagnetic field. As shown in figure 4.13, at plasmonic resonances, the scattering cross-section—and consequently the extinction—becomes much larger than the particle's geometric area. It is as if, at resonant wavelengths, the 'size' of the sail increases drastically.

Since the radiation force is proportional to the pressure (photon flux) times the area over which it acts, this increase in the cross-section leads to a significant amplification of the extinction force. For dielectric particles, although the photon flux is the same, the 'sail area'—that is, the extinction cross-section—is much smaller, resulting in a considerably weaker force.

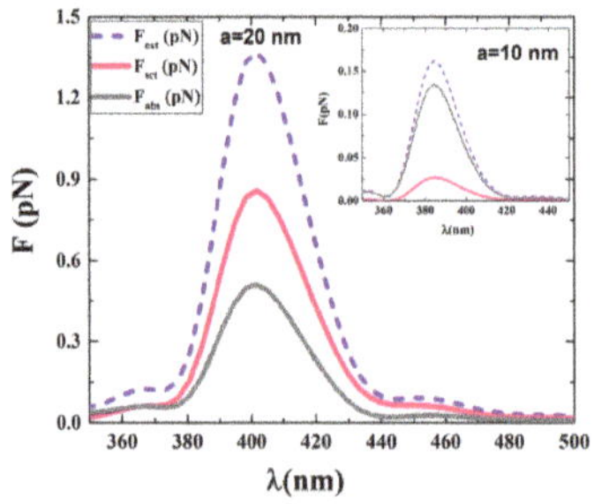

(a) Radiation pressure forces.

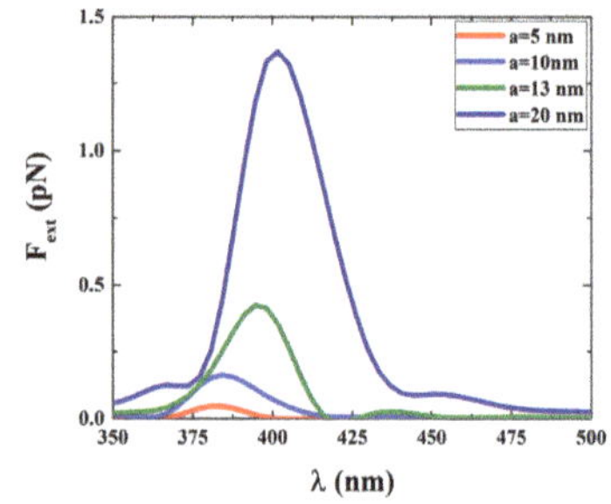

(b) Effect of AG NP size on extinction force.

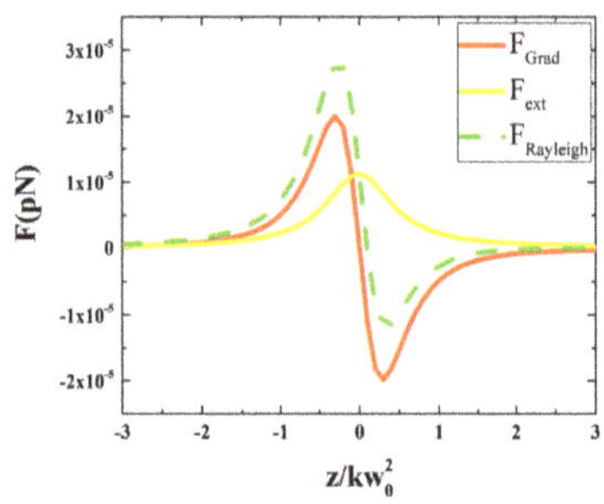

(c) Longitudinal forces for $\lambda = 1047$ nm and $a =$ 10 nm.

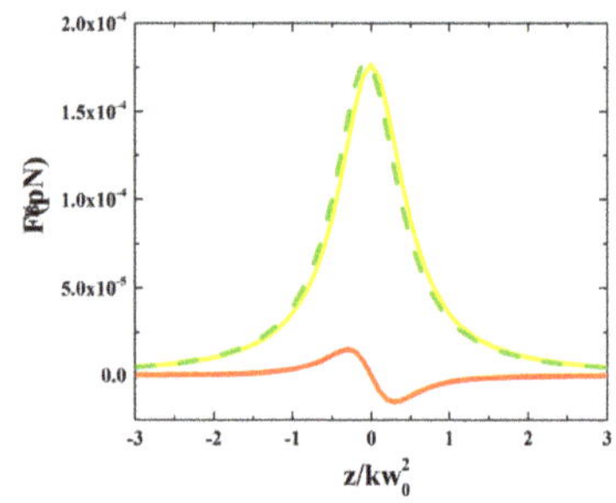

(d) [Longitudinal forces for $\lambda = 633$ nm and $a =$ 10 nm.

Figure 4.20. Longitudinal optical forces on AgNPs in water. (a) Radiation pressure forces on AgNPs as a function of λ with $a = 20$ nm. Inset: same for $a = 10$ nm. Solid gray line: $\mathbf{F}_{\text{abs}}$, solid pink line: $\mathbf{F}_{\text{sct}}$, dashed violet line: $\mathbf{F}_{\text{ext}}$. (b) Extinction force on AgNPs as a function of λ for different particle sizes. (c) Comparison of the longitudinal forces as a function of the normalized position z/kw_0^2 for $a = 10$ nm and $\lambda = 1047$ nm. (d) Same as (c) but for $a = 10$ nm and $\lambda = 633$ nm. Solid orange line: $\mathbf{F}_{\text{grad},z}$, solid yellow line: $\mathbf{F}_{\text{ext}}$, dashed green line: $\mathbf{F}_{\text{Rayleigh}} = \mathbf{F}_{\text{grad},z} + \mathbf{F}_{\text{ext}}$. Parameters: $w_0 = 2\ \mu$m, $P = 100$ mW, optical data for silver obtained from reference [17]. $\mathbf{r}_{\text{d}} = (0, 0, z/kw_0^2)$.

In other words, the force on plasmonic particles is amplified at resonance frequencies because the particle effective interaction area with light grows substantially due to the excitation of LSP resonance modes.

This does not mean that three-dimensional stable traps cannot be constructed with plasmonic particles. In addition to the extinction force, the gradient force also acts in the longitudinal direction and, just like the transverse component, is also affected by the LSP resonance.

In cases where the longitudinal gradient force is repulsive, it adds to the extinction force, preventing the formation of the trap. On the other hand, in situations where the gradient force is attractive, it opposes the extinction force, and trapping can be achieved if the gradient force is dominant.

This scenario is illustrated in figure 4.20(c), where we have a silver particle in water, with radius $a = 10$ nm, illuminated by a beam with $\lambda = 1047$ nm. Note that the gradient force (solid orange line) is slightly greater than the extinction force (solid yellow line), resulting in a total force, called the Rayleigh force (dashed green line, $\mathbf{F}_{\text{Rayleigh}} = \mathbf{F}_{\text{ext}} + \mathbf{F}_{\text{Grad},z}$), which acts as a restoring force, leading the particle to a stable equilibrium point near the focal plane.

A distinct scenario is shown in figure 4.20(d), where the same particle from the previous case is now illuminated by a beam with $\lambda = 633$ nm. Although, in this situation, the gradient force remains attractive, it becomes much smaller than the extinction force, causing the Rayleigh force on the particle to push it along the beam propagation direction.

For a plasmonic particle illuminated by a zero-order Gaussian beam, we can construct a criterion for the three-dimensional stability of the trap based on equation (2.94). Since only the gradient force acts in the transverse plane, the condition for transverse stability is that the gradient force is attractive.

Assuming the transverse stability condition is satisfied (i.e., the gradient force is attractive), the criterion for longitudinal trapping, and consequently for the overall stability of the trap, can be defined. For a stable 3D trap, the attractive longitudinal gradient force must be strong enough to overcome the repulsive extinction force. This requirement is elegantly captured by the parameter R:

$$R = \frac{F_{\text{Grad},z}\left(0,\, 0,\, \dfrac{kw_0^2}{2\sqrt{3}}\right)}{F_{\text{ext}}\left(0,\, 0,\, \dfrac{kw_0^2}{2\sqrt{3}}\right)} < -1 \tag{4.68}$$

where $(0,\, 0,\, \frac{kw_0^2}{2\sqrt{3}})$ is the position where the longitudinal gradient force reaches its maximum value. Here, the condition $R < -1$, simultaneously ensures that the gradient force is attractive ($F_{\text{Gradz}} < 0$, making R negative) and that its magnitude is greater than the extinction force ($|F_{\text{Gz}}| > |F_{\text{ext}}|$).

Note that this criterion is analogous to the one proposed by Ashkin [29] and Harada [30], with the difference that, in the case of plasmonic particles, we replace the scattering force with the extinction force. Note also that it was written to take into account the cases where the gradient force can be positive.

Figure 4.21 shows how the stability criterion varies with the particle radius for different wavelengths. Note that, for a beam with a waist of $w_0 = 2\ \mu$m, trapping a silver particle is only possible in the infrared (IR) region.

In fact, several authors have reported the construction of three-dimensional optical traps for plasmonic particles in the IR, such as Svoboda [23] and Hansen e*t al* [31], who extended the trapping range of gold particles to diameters up to 254 nm. Bosanac e*t al* [32] also achieved stable 3D trapping of silver particles at $\lambda = 1064$ nm, for radii ranging from 20 to 275nm, among other relevant works [25, 33].

The impossibility of constructing a 3D trap in the visible spectrum for silver particles stems directly from the LSP resonance. However, 2D traps have been reported in the literature in the visible range, as in the work of Furukawa and Yamaguchi [21], where gold particles with diameters between 0.5 and 3.0 μm were successfully trapped in two dimensions.

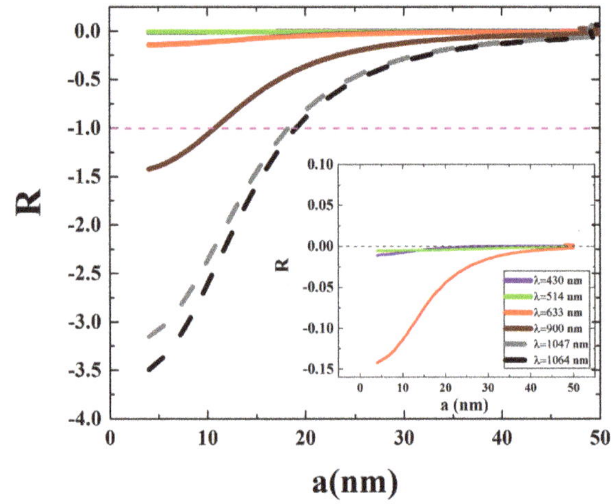

Figure 4.21. Axial stability criterion as a function of the radius for AgNPs in water at different illumination wavelengths. Parameters: $z = kw_0^2/2\sqrt{3}$, $w_0 = 2\ \mu$m, $P = 100$ mW, optical data for silver obtained from reference [17]. Solid violet line: $\lambda = 430$ nm, solid green line: $\lambda = 514$ nm, solid red line: $\lambda = 633$ nm, solid brown line: $\lambda = 900$ nm, dashed gray line: $\lambda = 1047$ nm, dashed black line: $\lambda = 1064$ nm. The dashed purple line marks the axis where $R = -1$.

Modifications to the trap configuration, such as reducing the beam waist or altering the intensity profile, have also been employed to achieve stable trapping both in the IR and in the visible range [25, 33,]. In the following section, we will present a method based on the use of surface plasmon polariton (SPP) fields for trapping metallic particles, enabling confinement over a wider range of particle sizes and wavelengths.

4.6 Plasmonic tweezers

The basic operation of traditional OT consists of exploiting the spatial variation of the optical beam intensity to induce an attractive force on a particle interacting with this beam. This force, called the gradient force, becomes stronger as the spatial variation of the field, i.e. its gradient (hence the name), increases, and as the beam intensity increases. Typically, OT are implemented by focusing the beam using a high-NA objective, which generates a significant gradient force at the focus. In the focal region, an optical potential well is formed, capable of trapping objects ranging in size from a few nanometers to several micrometers. Furthermore, the diffraction limit imposes a restriction on the minimum size of the focused beam, which is on the order of the wavelength of light, i.e. a few hundred nanometers. This limit not only defines the maximum achievable gradient force but also restricts the precision of traps based on this potential.

In this context, the so-called plasmonic optical tweezers (POT) have emerged, initially proposed to overcome the diffraction-imposed limitation on the focal spot size. However, POTs go beyond simply reducing this limit, as plasmonic structures enable an increase in the gradient force relative to extinction forces, allowing or enhancing the trapping of particles of various types, sizes, and materials. We will analyze this aspect in detail in chapter 6, where we will see that semiconductor particles can form stable 3D traps only in their plasmonic phase.

The basic principle of POT is to apply plasmonics concepts to the configuration of optical traps, combining both fields. In a way, all our discussion in the previous section already falls within this scope, since the concept of LSP resonance was used to improve the efficiency of optical traps. However, in this section, we will address another aspect of POT, in which plasmonics concepts are used not only to modify the particle's response but also to alter the optical beam itself. In particular, we will discuss how the generation of SPPs can be explored for constructing optical traps.

As discussed in section 4.2, at the interface between a dielectric medium and a conducting medium, it is possible to induce a propagating electromagnetic wave called an SPP. These waves have distinct characteristics from the incident wave that generates them. For instance, the SPP propagates with a specific wavenumber β (as well as other optical characteristics such as permittivity, refractive index, etc), as defined by equation (4.34), which implies that its wavelength is given by

$$\lambda_{\mathrm{SPP}} = \frac{2\pi}{\beta} = \lambda\,\mathrm{Re}\left\{\sqrt{\frac{\varepsilon_1 + \varepsilon_2}{\varepsilon_1 \varepsilon_2}}\right\}, \tag{4.69}$$

where λ is the vacuum wavelength of the beam, ε_1 is the permittivity of the dielectric medium, and ε_2 is the permittivity of the conducting medium. For example, if the SPP is generated at a water–gold interface with an incident beam of $\lambda = 514$ nm, we obtain $\lambda_{\mathrm{SPP}} = 329$ nm, assuming $\varepsilon_1 = 1.77$ and $\varepsilon_2 = -3.5 + 2.96i$. The SPPs propagate along the interface and decay exponentially in the dielectric and conductor, as described by the extinction factor γ_i (equation (4.36)), as illustrated in figure 4.5(b). The existence of these waves in the dielectric medium, with a reduced wavelength compared to the incident beam, is essential to understanding how plasmonic can overcome the diffraction-imposed focusing limit. In other words, by using the beam to excite an SPP and focusing it, we obtain a much smaller spot than that achieved by directly focusing the incident beam.

To excite SPP waves, coupling between the beam's wave vector and the SPP wave vector must occur, as illustrated in figure 4.6. A high-NA objective can be used for this purpose, functioning as a prism in the Kretschmann configuration [15].

In fact, the objective plays a crucial role in constructing the plasmonic tweezers, as it focuses the laser beam at the metal–dielectric interface, such as between gold and water. Using a high-NA objective, the beam is concentrated in a specific region, significantly increasing the intensity of the electromagnetic field. This intensity enhancement is essential for exciting SPPs, which require a sufficiently strong field to be generated. Additionally, the objective controls the beam's incidence angle, ensuring that it meets the interface under the necessary condition for coupling required to generate the SPP. This control allows for the creation of a well-defined focal point, essential for precise nanoparticles manipulation using the plasmonic tweezers.

The SPPs excited through the objective directly influence the optical forces by generating a highly confined and intensified electromagnetic field at the metal–

dielectric interface. When the beam is focused, it excites the SPPs, which propagate a highly localized evanescent field. This confined field generates a strong intensity gradient, amplifying the gradient force, which enables the precise capture and manipulation of nanoparticles. Furthermore, the extinction force, which arises from the momentum exchange between light and particles, is also intensified due to the high field confinement of the SPPs. However, in general, the strong field gradient makes the gradient force dominant.

In summary, the process unfolds as follows: a glass slide coated with a thin gold film is placed after the high-NA objective. The sample holder, containing suspended particles in water, is mounted on this thin gold film. The beam is focused by the objective and impinges on the glass/gold film interface. For the appropriate coupling angles, an SPP is generated at the gold/water interface, as illustrated in figure 4.6 and described by equation (4.39). The excitation of SPPs at the gold/water interface, whose propagation and characteristics can be visualized through the detection of their leakage radiation via the substrate (as illustrated for an analogous system in figure 4.8), results in the formation of an evanescent, highly confined, and intense electromagnetic field at the interface region on the water side. It is this SPP field at the gold/water interface that generates the significant optical field gradient. This strong gradient results in a dominant gradient force over the scattering force, creating more defined (smaller spot) and more efficient traps.

The construction of plasmonic tweezers enables the manipulation of particles that would not be possible in traditional setups, such as the trapping of metallic particles at microscopic scales. A notable example in this application is the work of Min *et al* [24], in which they successfully manipulated gold particles with diameters ranging from (0.2–2) μm that could not be trapped in a standard OT configuration. Min *et al*'s work presents an additional advantage over other plasmonic tweezers setups [34–36]: in their configuration, not only does the gradient force dominate the scattering force, but the scattering force itself is also utilized as an auxiliary force in the construction of the trap.

To understand this, let us examine the work of Min *et al* in detail. Figure 4.22(a) illustrates the setup used by Min, which consists of the incidence of a radially polarized light beam ($\lambda = 1064$ nm) on a high-NA objective ($NA = 1.49$) focused onto a glass substrate coated with a thin gold film. As the beam passes through a series of optical components, including polarizers and phase plates, it is transformed into a radially polarized beam that, upon reaching the Au/water interface, excites SPPs at the resonance angle. These SPPs propagate circularly towards the center of the interface, where they interfere constructively, forming a highly focused intensity peak known as the 'SPP virtual probe'. This peak creates a strong and localized electromagnetic field that attracts and traps metallic particles (figure 4.22(b)), leveraging the enhanced gradient force and the aligned scattering force, allowing for the efficient manipulation of metallic particles of different sizes. The theoretical full width at half maximum of the virtual probe is 261 nm ($\sim 0.245\lambda$) with $\lambda_{SPP} = 754$ nm.

The gradient force is induced on the metallic particle due to the interaction between the highly focused electric field of the SPPs and the particle. When the metallic particle is exposed to the intense and localized electric field generated by the

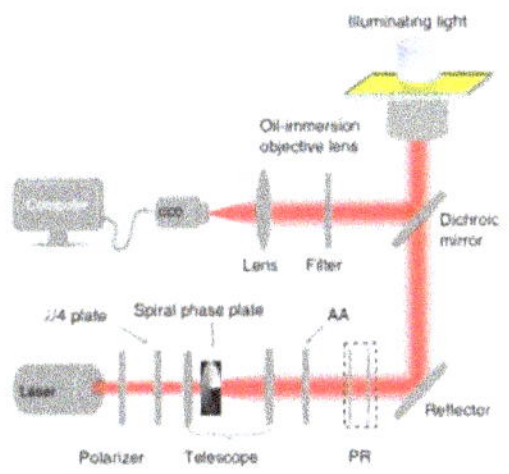

(a) Experimental setup of the focused plasmonic trapping system.

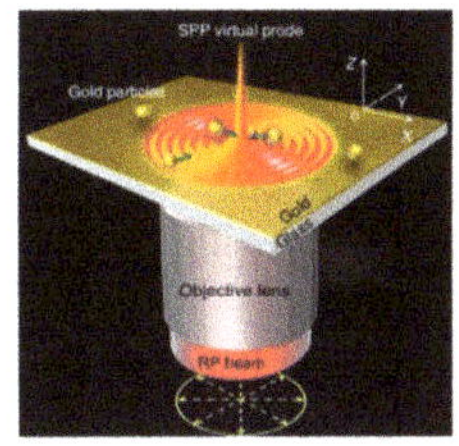

(b) Schematic of trapping metallic particles by a SPP virtual probe

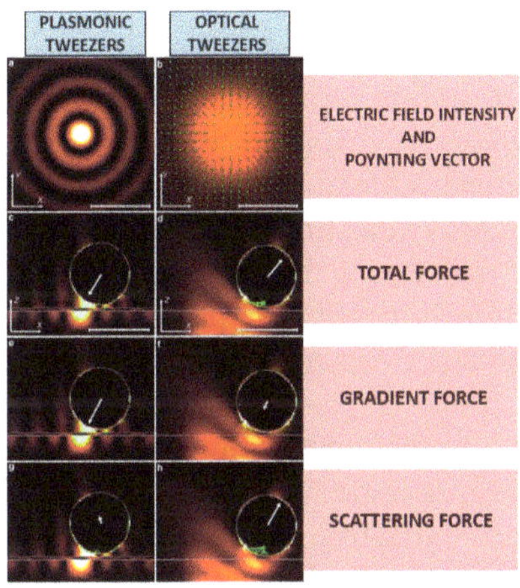

(c) Comparison of forces in plasmonic and optical tweezers.

Figure 4.22. Plasmonic tweezers setup developed by Min *et al.* (a) Experimental setup of the focused plasmonic trapping system. The incident wavelength is $\lambda = 1{,}064$ nm (vacuum), the thickness of the thin gold film is 45 nm, the refractive index of the glass substrate is 1.515, and the diameter of the gold particles is 1 ± 0.1 μm. The theoretical full width at half maximum of the SPP virtual probe is 261 nm ($\sim 0.245\lambda$). (b) Schematic of trapping metallic particles by an SPP virtual probe. The bottom yellow arrows indicate the polarized directions of the radially polarized (RP) beam; the blue arrows indicate the direction of force on each gold particle in the SPP virtual probe field. (c) Distribution of electric field intensity (background) and Poynting vector (green arrows) in the horizontal x–y plane of (a) the focused plasmonic tweezers and (b) the focused OT, where the x–y plane is 50 nm above the gold–water interface in (a) and the glass–water interface in (b), respectively. (c) and (d) show the total force, (e) and (f) the gradient force, and (g) and (h) the scattering force, respectively (green arrows), distributed on a gold particle (diameter of 1 μm) in the vertical x–y plane for the plasmonic tweezers (c, e, g) and the OT (d, f, h). The background is the electric field intensity, while the white lines indicate the spherical particle and gold film. The particle with a diameter of 1 μm is 50 nm above the gold surface and 300 nm away from the SPP peak at the horizontal center for plasmonic tweezers, while in OT, it is 600 nm away from the horizontal center. The white arrow starting from the center of the sphere in (c–h) denotes the resultant force on the particle. The length of all scale bars (white line in the lower right corner) is 1 μm. Reproduced from [24], with permission from Springer Nature.

plasmonic tweezer, charges are induced on its surface (LSP), creating a Coulomb force that depends on the intensity and distribution of the local electric field. This gradient force pulls the particle towards the peak intensity of the electric field, allowing its capture and manipulation.

As in the traditional case, the scattering force is proportional to the scattering cross-section and the average Poynting vector. In this configuration, the Poynting vector leads to an attractive scattering force due to the way SPPs are generated and propagate. When the radially polarized light beam is incident on the Au/water interface, it excites SPPs that propagate towards the center, forming a focused and intense field. In this process, the SPPs act as secondary circular sources propagating inward, creating constructive interference that results in an intensity peak at the center. The Poynting vector, which represents the flow of electromagnetic energy, points in the direction of the SPP propagation. Since the SPPs move towards the center, the Poynting vector also points towards the center, aligning with the gradient force.

Figure 4.22(c) compares the distributions of the electric field, Poynting vectors, and total, gradient, and scattering forces in plasmonic tweezers (left) and traditional OT (right), where the forces are evaluated for a gold particle of 1 μm in diameter, obtained through numerical simulations using the finite-difference time-domain (FDTD) method and the Maxwell stress tensor (MST) method. Figures (a) and (b) show the distribution of the electric field intensity (background) and the Poynting vectors (green arrows) in the horizontal xy-plane above the water–gold (plasmonic) and water–glass (traditional) interfaces, respectively, where the central peak is stronger and more concentrated in plasmonic tweezers, with Poynting vectors nearly disappearing, indicating a lower power flow. Figures (c) and (d) present the total force distribution on a gold particle in the vertical plane (xy-plane), with the total force in plasmonic tweezers indicating a horizontal direction to the left, while in OT, it indicates a horizontal direction to the right. Figures (e) and (f) show the gradient force that attracts particles to the center, which is much stronger in plasmonic tweezers. Figures (g) and (h) illustrate the scattering force, which in OT opposes the gradient force, pushing the particle away from the center, whereas in plasmonic tweezers, it assists the gradient force, with both acting in the same direction to attract the particle to the center. The green arrows in figures (c)–(h) indicate the direction of the respective force.

While in plasmonic tweezers the gold particle is trapped, in the conventional tweezers it is pushed away from the beam spot. In addition to capturing metallic particles, plasmonic tweezers enhance the trapping efficiency of dielectric particles and Rayleigh metallic particles. Another application gaining traction in the field of plasmonic tweezers is their potential use in surface-enhanced Raman scattering [37, 38], strongly suggesting that the implementation of plasmonic in the field of optical manipulation holds a promising future.

References

[1] Maier S A 2007 *Plasmonics: Fundamentals and Applications.* (Berlin: Springer)

[2] Marder M P 2010 *Condensed Matter Physics.* (New York: Wiley)

[3] Kittel C and McEuen P 2018 *Introduction to Solid State Physics.* (New York: Wiley)
Saleh B E A and Teich M C 2019 *Fundamentals of Photonics* (New York: Wiley)

[4] Zeman E J S G C 2011 An accurate electromagnetic theory study of surface enhancement factors for silver, gold, copper, lithium, sodium, aluminum, gallium, indium, zinc, and cadmium *J. Phys. Chem.-US* **91** 634
[5] Pettit R B, Silcox J and Vincent R 1975 Measurement of surface-plasmon dispersion in oxidized aluminum films *Phys. Rev.* B **11** 3116
[6] Vincent R and Silcox J 1973 Dispersion of radiative surface plasmons in aluminum films by electron scattering *Phys. Rev. Lett.* **31** 1487
[7] Gong S, Hu M, Zhong R, Chen X, Zhang P, Zhao T and Liu S 2014 Electron beam excitation of surface plasmon polaritons *Opt. Exp.* **22** 19252
[8] Devaux E, Ebbesen T W, Weeber J-C and Dereux A 2003 Launching and decoupling surface plasmons via micro-gratings *Appl. Phys. Lett.* **83** 4936
[9] Huang M, Zhao F, Cheng Y, Xu N and Xu Z 2009 Mechanisms of ultrafast laser-induced deep-subwavelength gratings on graphite and diamond *Phys. Rev. B Cond. Matter Mater. Phys.* **79** 125436
[10] Bouhelier A and Wiederrecht G P 2005 Surface plasmon rainbow jets *Opt. Lett.* **30** 884
[11] Hecht B, Bielefeldt H, Novotny L, Inouye Y and Pohl D W 1996 Local excitation, scattering, and interference of surface plasmons *Phys. Rev. Lett.* **77** 1889
[12] Omar N A S and Fen Y W 2018 Recent development of spr spectroscopy as potential method for diagnosis of dengue virus e-protein *Sensor Rev.* **38** 106
[13] Steglich P, Lecci G and Mai A 2022 Surface plasmon resonance (SPR) spectroscopy and photonic integrated circuit (PIC) biosensors: a comparative review *Sensors* **22** 2901
[14] Wing Fen Y and Mahmood Mat Yunus W 2013 Surface plasmon resonance spectroscopy as an alternative for sensing heavy metal ions: a review *Sens. Rev.* **33** 305
[15] Kretschmann E and Raether H 1968 Radiative decay of non radiative surface plasmons excited by light *Z. Naturforsch.* A **23** 21356
[16] Otto A 1968 Excitation of nonradiative surface plasma waves in silver by the method of frustrated total reflection *Z. Phys. A: Hadrons Nucl.* **216** 398
Geonmonond R S, da Silva A G M, Rodrigues T S, de Freitas I C, Ando R A, Alves T V and Camargo P H C 2018 Addressing the effects of size-dependent absorption, scattering, and near-field enhancements in plasmonic catalysis *ChemCatChem.* **10** 3447
[17] Johnson P B and Christy R-WJPrB 1972 Optical constants of the noble metals *Phys. Rev.* B **6** 4370
[18] Mock J J, Barbic M, Smith D R, Schultz D A and Schultz S h 2002 Shape effects in plasmon resonance of individual colloidal silver nanoparticles *J Chem. Phys.* **116** 6755
[19] Bohren C F and Huffman D R 2008 *Absorption and scattering of light by small particles* (New York: Wiley)
[20] Sonnefraud Y, Koh A L, McComb D W and Maier S A 2012 Nanoplasmonics: engineering and observation of localized plasmon modes *Laser Photon. Rev.* **6** 277
[21] Furukawa H and Yamaguchi I 1998 Optical trapping of metallic particles by a fixed Gaussian beam *Opt. Lett.* **23** 216
[22] Chaumet P C and Nieto-Vesperinas M 2000 Electromagnetic force on a metallic particle in the presence of a dielectric surface *Phys. Rev.* B **62** 11185
[23] Svoboda K and Block S M 1994 Optical trapping of metallic rayleigh particles *Opt. Lett.* **19** 930–2
[24] Min C, Shen Z, Shen J, Zhang Y, Fang H, Yuan G, Du L, Zhu S, Lei T and Yuan X 2013 Focused plasmonic trapping of metallic particles *Nat. Commun.* **4** 2891

[25] Lehmuskero A, Johansson P, Rubinsztein-Dunlop H, Tong L and Kall M 2015 Laser trapping of colloidal metal nanoparticles *ACS Nano* **9** 3453
[26] Rocha M S 2023 *DNA Interactions with Drugs and Other Small Ligands - Single Molecule Approaches and Techniques* 1st edn (New York: Academic)
[27] Oliveira L, Caquito J M and Rocha M S 2018 Carboplatin as an alternative to cisplatin in chemotherapies: New insights at single molecule level *Biophys. Chem.* **241** 8–14
[28] Bazoni R F, Moura T A and Rocha M S 2020 Hydroxychloroquine exhibits a strong complex interaction with DNA: unraveling the mechanism of action *J. Phys. Chem. Lett.* **11** 9528
[29] Ashkin A, Dziedzic J M, Bjorkholm J E and Chu S 1986 Observation of a single-beam gradient force optical trap for dielectric particles *Opt. Lett.* **11** 288
[30] Harada Y and Asakura T 1996 Radiation forces on a dielectric sphere in the rayleigh scattering regime *Opt. Commun.* **124** 529
[31] Hansen P M, Bhatia V K, Harrit N and Oddershede L 2005 Expanding the optical trapping range of gold nanoparticles *Nano Lett.* **5** 1937
[32] Bosanac L, Aabo T, Bendix P M and Oddershede L B 2008 Efficient optical trapping and visualization of silver nanoparticles *Nano Lett.* **8** 1486
[33] Ricardo Arias-González J and Nieto-Vesperinas M 2003 Optical forces on small particles: attractive and repulsive nature and plasmon-resonance conditions *J. Opt. Soc. Am.* A **20** 1201
[34] Zhang Y, Min C, Dou X, Wang X, Paul Urbach H, Somekh M G and Yuan X 2021 Plasmonic tweezers: for nanoscale optical trapping and beyond *Light: Sci. Appl.* **10** 59
[35] Ren Y, Chen Q, He M, Zhang X, Qi H and Yan Y 2021 Plasmonic optical tweezers for particle manipulation: principles, methods, and applications *ACS Nano* **15** 6105
[36] Juan M L, Righini M and Quidant R 2011 Plasmon nano-optical tweezers *Nat. Photon.* **5** 349
[37] Braun G, Lee S J, Dante M, Nguyen T-Q, Moskovits M and Reich N 2007 Surface-enhanced raman spectroscopy for dna detection by nanoparticle assembly onto smooth metal films *J. Am. Chem. Soc.* **129** 6378
[38] Du L, Yuan G, Tang D and Yuan C 2011 Tightly focused radially polarized beam for propagating surface plasmon-assisted gap-mode raman spectroscopy *Plasmonics* **6** 651

IOP Publishing

Optical Trapping and Manipulation of New Materials

Tiago de Assis Moura, Joaquim Bonfim Santos Mendes and Márcio Santos Rocha

Chapter 5

Semiconductor optics

After presenting a robust discussion on the optical trapping and manipulation of dielectric and metallic microparticles; in this chapter, we will briefly discuss the optical properties of semiconductors—materials that exhibit intermediate properties between dielectrics and conductors. Such discussion is important to understand how these properties play a role on the strong competition between the combined forces/effects that will appear when one intends to trap a semiconductor bead using optical tweezers; a topic that will be developed in the following chapters.

5.1 Plasma oscillations

Let us consider a piece of metal (or semiconductor, which also has free-charge carriers) as a gas of free electrons and a backdrop of positive ions. In equilibrium, free electrons and positive ions uniformly occupy the entire space, as illustrated in figure 5.1(a).

Let N represent the electron density at equilibrium, which must also be equal to the positive ion density, since the plasma is electrically neutral. Suppose that during a finite time interval t' we excite the plasma, causing the electrons to move from their equilibrium position. Since the mass of the ions is much greater than that of the electrons, we can consider the electrons to be moving entities, while the ions remain fixed. Consequently, there is an increase in the density of electrons (negative charge) in the region to which they migrated and an increase in the density of ions (positive charges) in the region from which the electrons migrated.

Under these conditions, a restoring electric field emerges in the region between the charge densities, aiming to bring the system back to equilibrium, as illustrated in figure 5.1(b). Neglecting dissipation, which inevitably occurs in real systems, this system will sustain oscillations of electron charges in a simple harmonic motion. If we include dissipation, the electron charge density returns to the equilibrium value, executing a damped harmonic motion. In the case of a sub-critical system, the charge density will oscillate a few times before returning to its equilibrium value. The

doi:10.1088/978-0-7503-6074-6ch5

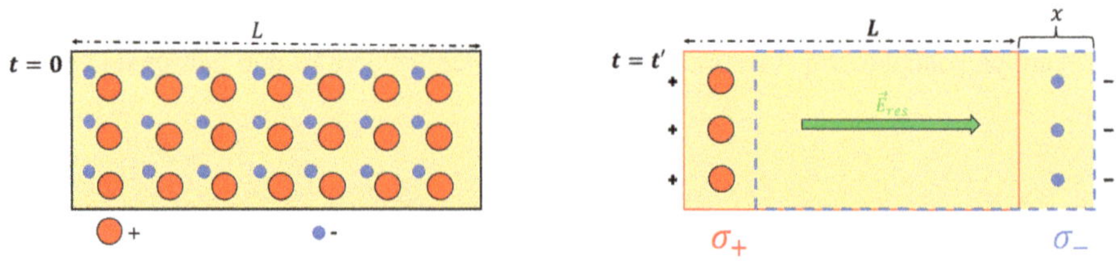

(a) In equilibrium. (b) In the presence of an external electric field.

Figure 5.1. Scheme illustrating the origin of plasma oscillations. (a) Free electron gas in equilibrium at $t = 0$. (b) Separation of positive and negative charges due to the disturbance of an external electric field and the emergence of the restoring electric field ($\mathbf{E}_{\text{res}}$) after the disturbance.

interesting problem is to calculate how the charge density inside a conductor approaches zero as the conductor tends to electrostatic equilibrium, as demonstrated in the work of Bochove *et al* [1].

Let x be the net displacement of the free electrons. Note that, physically, our problem is equivalent to that of a parallel plate capacitor. By approximating the restoring electric field ($\mathbf{E}_{\text{res}}$) to that of an electric field in a parallel plate capacitor, we can write

$$E_{\text{res}} = \frac{\sigma}{\varepsilon_0} = \frac{Nex}{\varepsilon_0}, \tag{5.1}$$

where $\sigma = Nex$ is the surface charge density, N is the electron density, and e the electron charge.

Then, the restoring force acting on the electrons is (neglecting damping):

$$F_{\text{res}} = -eE_{\text{res}} = -\frac{Ne^2x}{\varepsilon_0}. \tag{5.2}$$

The equation of motion for the electrons (neglecting damping) can be written as

$$m_e\partial_t^2x(t) = -\frac{Ne^2x(t)}{\varepsilon_0} \rightarrow \partial_t^2x(t) + \omega_p^2x(t) = 0, \tag{5.3}$$

whose general solution is

$$x(t) = A\cos(\omega_p t + \delta), \tag{5.4}$$

where ω_p is the frequency of the oscillations, also called plasma frequency, given by

$$\omega_p = \sqrt{\frac{Ne^2}{m_e\varepsilon_0}}. \tag{5.5}$$

Note that the disturbance in the plasma will induce free oscillations of the electrons around their equilibrium positions with a natural frequency ω_p, which is proportional to the square root of the electron density. In other words, ω_p is related to the response time of electrons within the plasma when it is disturbed.

An interesting scenario occurs when the plasma is disturbed by an electromagnetic (EM) wave with frequency ω_{EM}. If $\omega_{EM} \gg \omega_{\mathrm{p}}$, then the response time of the electrons is much greater than the variation time of the electric field. As a result, for the EM wave, the plasma behaves as if it were transparent.

The plasma frequency is related to various notable effects. For instance, to transmit radio waves through the ionosphere, the wave frequency must exceed ω_{p}; otherwise, the signal will be reflected back. Conversely, for communication with a radio station beyond the horizon, using frequencies below ω_{p} ensures that the signal is reflected in the ionosphere and directed to the receiver.

5.2 The Drude model

In the beginning of the 20th century, Paul Drude proposed that, in metals, the electrons in the outermost shell of the atoms can become free, while the ions remain fixed. The Drude model involves considering a gas of electrons and applying the kinetic theory of gases to simplify the problem. In the context of this model, transport is facilitated by the free electrons. When an electric field is applied, it generates an electric current, and through collisions between free electrons and the fixed ions of the crystal lattice, the electrons transfer momentum to the lattice ions. The key ideas of the Drude Model can be summarized in three hypotheses:

1. The electrons are independent (electron–electron interaction is neglected). Also, electrons are free (electron–lattice interactions are neglected except during the instants when electrons collide with the ions of the lattice).
2. Collisions between electrons and ions are modeled as instantaneous events that occur with a probability per unit of time given by $\frac{1}{\tau_{\mathrm{c}}}$.
3. Electrons thermalize with their surroundings only through collisions. We assume that the collisions are isotropic, so that after a collision we will have $\langle \vec{v} \rangle = 0$ and $\frac{m}{2}\langle v^2 \rangle = \frac{3K_{\mathrm{b}}T}{2}$.

Based on these hypotheses we can express the momentum of the electron after an instant of time dt as:

$$\mathbf{p}(t) = \begin{cases} \mathbf{p}_{\mathrm{col}}, & \text{if there is a collision} \\ \mathbf{p}(t) + \mathbf{F}dt, & \text{if there is no collision} \end{cases} \tag{5.6}$$

where $\mathbf{p}(t)$ is the initial momentum, $\mathbf{p}_{\mathrm{col}}$ is the momentum due to the collision and $\mathbf{F}dt$ is the impulse of the external force.

Based on the second hypothesis we can assume that:

- The probability of an electron to experiment a collision in the interval between t and $t + dt$ is $\frac{dt}{\tau_{\mathrm{c}}}$.
- Therefore, the probability of an electron not colliding in such interval is $(1 - \frac{dt}{\tau_{\mathrm{c}}})$.

Thus, we rewrite equation (5.6) as

$$\mathbf{p}(t + dt) = \frac{dt}{\tau_c}\mathbf{p}_{col} + \left(1 - \frac{dt}{\tau_c}\right)(\mathbf{p}(t) + \mathbf{F}dt). \tag{5.7}$$

Taking the average over n electrons, we write

$$\langle \mathbf{p}(t + dt)\rangle = \frac{dt}{\tau_c}\langle \mathbf{p}_{col}\rangle + \left(1 - \frac{dt}{\tau_c}\right)(\langle \mathbf{p}(t)\rangle + \mathbf{F}dt). \tag{5.8}$$

According to the third hypothesis, collisions are isotropic; therefore, $\langle \mathbf{p}_{col}\rangle = 0$, such that

$$\langle \mathbf{p}(t + dt)\rangle = \left(1 - \frac{dt}{\tau_c}\right)(\langle \mathbf{p}(t)\rangle + \mathbf{F}dt), \tag{5.9}$$

which implies

$$\partial_t\langle \mathbf{p}(t)\rangle = -\frac{\langle \mathbf{p}(t)\rangle}{\tau_c} + \mathbf{F}. \tag{5.10}$$

The left side is the dynamics term, while in the right the first term represents the dissipation term and the second term denotes the external force applied to the electrons in the metal.

5.3 Optics of metals and semiconductors

Metals and semiconductors, which have excess free-charge carriers, can be effectively treated as a gas plasma. This is due to the presence of equal numbers of fixed positive ions and free electrons (or free electrons and holes in the case of intrinsic semiconductors) within these materials. Notably, free electrons in metals and semiconductors experience restoring forces when interacting with EM waves. In the case of semiconductors, the contribution of bound electrons becomes relevant, and their impact can be calculated using the Lorentz model.

Let us delve into the oscillations of free electrons in metals, induced by an EM wave. The field has an angular frequency ω and amplitude E_0, polarized along the x-axis direction. The equation for electron displacement, denoted as x, can be derived from the Drude model. Utilizing equation (5.10) we obtain

$$\partial_t^2 x(t) + \gamma\partial_t x(t) = -\frac{-e}{m_e}E(t) = \frac{-e}{m_e}E_0 e^{-i\omega t}, \tag{5.11}$$

where $\gamma = \frac{1}{\tau_a}$ is the damping rate of free electrons. The first term in the equation represents the acceleration of the electron, while the second term represents the frictional damping force with the medium. On the right side, the term denotes the driving force exerted by the incident light. The particular (stationary) solution of (5.11) is

$$x(t) = X_0 e^{-i\omega t}. \tag{5.12}$$

Replacing equation (5.12) into equation (5.11), we write

$$x(t) = \frac{eE}{m_e(\omega^2 + i\gamma\omega)}. \tag{5.13}$$

The modulus of the polarization vector is denoted as P and is given by $P = -Nex$, where N is the number of electrons per unit volume. To determine the dielectric constant (ε_r) of the electron gas, we turn to the definition of the electrical displacement (D):

$$\begin{aligned} D &= \varepsilon_r \varepsilon_0 E \\ &= \varepsilon_0 E + P \\ &= \varepsilon_0 E - \frac{Ne^2 E}{m_e(\omega^2 + i\gamma\omega)}. \end{aligned} \tag{5.14}$$

Note that we can write ε_r as

$$\varepsilon_r(\omega) = 1 - \frac{Ne^2}{\varepsilon_0 m_e} \frac{1}{(\omega^2 + i\gamma\omega)}. \tag{5.15}$$

Using equation (5.5), we can rewrite equation (5.15) as

$$\varepsilon_r(\omega) = 1 - \frac{\omega_p^2}{(\omega^2 + i\gamma\omega)}. \tag{5.16}$$

We can apply the free electron model to semiconductors by implementing two pertinent modifications:

1. We need to account for the fact that, in semiconductors, electrons and holes move in the conduction and valence bands, respectively. This can be easily addressed by assuming that the charge carriers behave like particles with an effective mass m^*.
2. Semiconductors have bound electrons that contribute to the permittivity of the material. An effective way to consider bound electrons is to divide the polarizability $\mathbf{P}$ of the material into two parts: one resulting from the free-charge carriers ($\mathbf{P}_{free}$) and another resulting from the bound electrons ($\mathbf{P}_{bound}$).

Therefore,

$$D = \varepsilon_r \varepsilon_0 E, \tag{5.17}$$

$$D = \varepsilon_0 E + P_{bound} + P_{free}, \tag{5.18}$$

$$D = \varepsilon_0 \varepsilon_{opt} E - \frac{Ne^2 E}{m^*(\omega^2 + i\gamma\omega)}. \tag{5.19}$$

The first term in equation (5.19) represents the contribution of bound electrons to the electrical displacement. Therefore, $\varepsilon_{\text{opt}} = \varepsilon_{\text{opt}}' + i\varepsilon_{\text{opt}}''$ is the intrinsic dielectric constant of the semiconductor. The second term represents the contribution of free-charge carriers, and therefore N is the density of free carriers (electrons and holes).

We can write the dielectric constant as a function of frequency as

$$\varepsilon_r(\omega) = \varepsilon_{\text{opt}}' - \frac{Ne^2}{\varepsilon_0 m^*} \frac{1}{(\omega^2 + i\gamma\omega)} + i\varepsilon_{\text{opt}}'',$$

$$\varepsilon_r(\omega) = \varepsilon_{\text{opt}}' \left(1 - \frac{\omega_{\text{p}}^2}{(\omega^2 + i\gamma\omega)}\right) + i\varepsilon_{\text{opt}}''. \tag{5.20}$$

Note that now the plasma frequency ω_{p} is given by

$$\omega_{\text{p}}^2 = \frac{Ne^2}{\varepsilon_{\text{opt}}' \varepsilon_0 m^*}. \tag{5.21}$$

The physical interpretation of the plasma frequency for a semiconductor differs slightly from that for a free electron gas. Now, ω_{p} denotes the response time of both (free) electrons and holes when they are disturbed by an external field. Since we have two types of carriers, we work with $m^* = (\frac{1}{m_e^*} + \frac{1}{m_b^*})^{-1}$, which is the reduced effective mass. Here, m_{e}^* is the effective mass of the electron, and m_{b}^* is the effective mass of the hole. Additionally, we include the term ε_{opt} to account for polarizability due to bound electrons.

Rewriting $\varepsilon_r(\omega) = \varepsilon_1(\omega) + i \quad \varepsilon_2(\omega)$, we have

$$\varepsilon_1(\omega) = \varepsilon_{\text{opt}}' \left(1 - \frac{\omega_{\text{p}}^2 \tau_{\text{a}}^2}{\omega^2 \tau_{\text{a}}^2 + 1}\right), \tag{5.22}$$

$$\varepsilon_2(\omega) = \varepsilon_{\text{opt}}' \left(\frac{\omega_{\text{p}}^2 \tau_{\text{a}}}{\omega(1 + \tau_{\text{a}}^2 \omega^2)}\right) + \varepsilon_{\text{opt}}''. \tag{5.23}$$

The complex refractive index of the semiconductor, $\tilde{n} = n + i\kappa$, can be calculated using the components of the dielectric constant through the following relations:

$$n = \frac{1}{\sqrt{2}}\left[\varepsilon_1 + \left(\varepsilon_1^2 + \varepsilon_1^2\right)^{\frac{1}{2}}\right]^{\frac{1}{2}}, \tag{5.24}$$

$$\kappa = \frac{1}{\sqrt{2}}\left[-\varepsilon_1 + \left(\varepsilon_1^2 + \varepsilon_1^2\right)^{\frac{1}{2}}\right]^{\frac{1}{2}}. \tag{5.25}$$

The reflectivity R can be expressed as a function of the refractive index using the Fresnel equation

$$R(\tilde{n}) = \left| \frac{\tilde{n}-1}{\tilde{n}+1} \right|^2. \tag{5.26}$$

Unlike a metal, the charge carrier density of a semiconductor can be easily modified, as we will see in the next section. Since the plasma frequency varies with the square root of the carrier density, we can readily adjust the plasma frequency of semiconductors, thereby influencing their entire optical response. Figure 5.2 illustrates this scenario (here $\varepsilon_{\text{opt}}'' = 0$), depicting the variation of n, κ, and R for a semiconductor with three different carrier densities. These densities are represented by the plasma wavelength λ_{p}, and the variation is shown as a function of the excitation wavelength $\lambda_{\text{Excitation}}$. The plasma wavelength is the length associated with the plasma frequency, calculated by the usual relationship $\omega_{\text{p}} = \frac{2\pi c}{\lambda_{\text{p}}}$.

Note that the variation in the charge carrier density modifies the entire optical response of the material. For a specific wavelength (less than λ_{p}), an increase in the

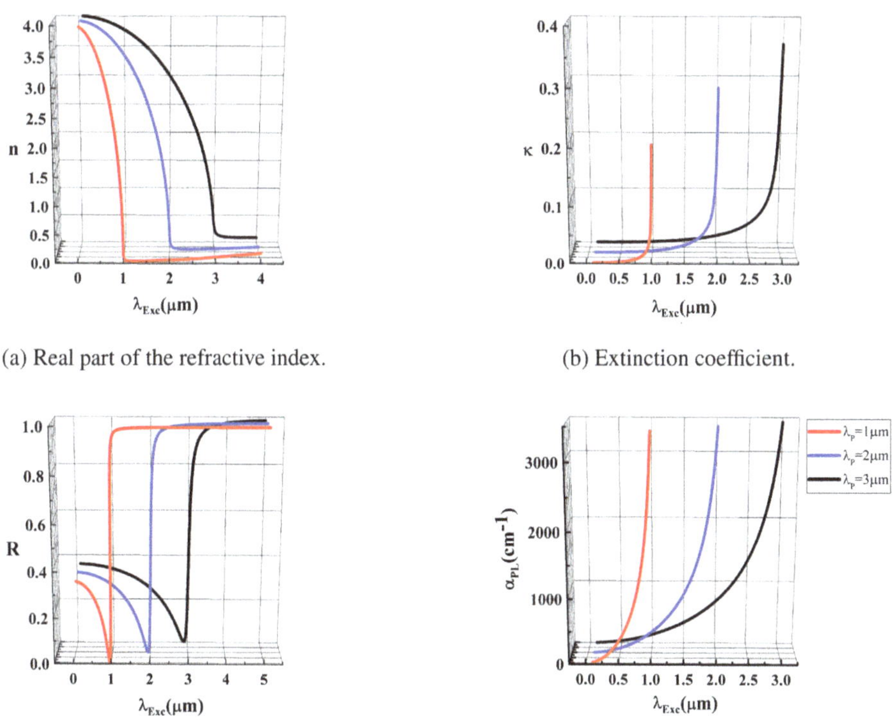

(a) Real part of the refractive index.

(b) Extinction coefficient.

(c) Reflectance.

(d) Free carrier absorption coefficient.

Figure 5.2. Variation of the optical properties of semiconductors as a function of the excitation wavelength. *Red solid line:* $\lambda_{\text{p}} = 1\ \mu\text{m}$. *Blue solid line:* $\lambda_{\text{p}} = 2\ \mu\text{m}$. *Black solid line:* $\lambda_{\text{p}} = 3\ \mu\text{m}$. (a) Variation of the real part of the refractive index. (b) Variation in the extinction coefficient. (c) Variation in reflectance. (d) Variation in the absorption coefficient due to free carriers. *Other parameters:* $\varepsilon_{\text{opt}}' = 16$; $\varepsilon_{\text{opt}}'' = 0$; $\tau_{\text{a}} = 10^{-13}$ s.

carrier density (resulting in a decrease in λ_p) leads to a simultaneous decrease in both n and R, while κ increases. When $\lambda = \lambda_p$, resonance occurs between the radiation frequency and the plasma frequency, significantly increasing the value of κ, while $n \cong 0$ and $R \cong 0$. For values of $\lambda > \lambda_p$, the radiation is strongly absorbed by free carriers, characterized by the fact that $\kappa \gg n$, implying that $R \cong 1$.

The imaginary part of the refractive index is associated with the absorption of EM waves. The relationship between α and κ can be derived by considering the propagation of an EM wave in a medium with a complex refractive index $\tilde{n}$. As the wave penetrates the material, the intensity of the EM wave will decay exponentially due to κ. By comparing the decay of the intensity of the EM wave (expressed in terms of the complex refractive index) with Beer's law, we obtain the following relationship

$$\alpha = \frac{4\pi}{\lambda}\kappa. \tag{5.27}$$

Equation (5.27) is valid for all absorption mechanisms that we will discuss in the next section. A significant advantage of discerning α is the ability to understand how each mechanism works and impacts semiconductor optics. In figure 5.2(d), we illustrate the variation of α_{FC} (FC = free carriers) described by equation (5.27) for a semiconductor with three different carrier densities represented by λ_p. The variation of κ is given by equation (5.25). As expected, increasing the carrier density raises the value of α_{FC}. Furthermore, for $\lambda \geqslant \lambda_P$, the value of α_{FC} increases significantly due to plasma resonance.

5.3.1 Carrier generation through optical absorption in semiconductors

In addition to thermal excitation [2, 3], semiconductors can be excited through optical absorption. Electron excitation occurs when the photon energy ($\hbar\omega$) is sufficient to cause the electron to 'jump' from a state in the valence band to a state in the conduction band ($\hbar\omega > E_{gap}$). This transition can occur directly, without a change in the electron's wave vector (also called vertical transitions), or it can occur indirectly. In indirect transitions, the electron changes its wave vector, which, due to the conservation of linear momentum, requires the absorption (or emission) of a phonon in addition to the absorption of a photon. In other words, if the minimum of the conduction band and the maximum of the valence band occur at the same 'crystal momentum'[1], the semiconductor is said to be a direct gap semiconductor, and the transitions that occur are direct transitions. However, if the minimum of the conduction band and the maximum of the valence band occur over different crystal momenta, then the semiconductor is said to be an indirect gap semiconductor, and the transitions that occur are indirect transitions.

Indirect transitions occur because photons have energy comparable to the distance between the bands but negligible momentum compared to the Brillouin

[1] Crystal momentum is the quasi-momentum (or momentum-like vector) associated with electrons and holes in the crystal lattice.

zone, whereas phonons have momentum compatible with the Brillouin zone but have negligible energy compared to the distance between the bands. Consequently, indirect transitions have a much lower probability of occurring compared to direct transitions, which do not require the assistance of phonons.

In many situations of interest, the photon has enough energy (E_{photon}) to carry out a direct transition, even if the semiconductor is an indirect gap semiconductor. This is why many authors use the nomenclature 'direct gap' ($E_{\text{gap}}^{\text{D}}$) to refer to the minimum energy required for a vertical transition between semiconductor bands to occur. A notable example is germanium, which at 300 K has an indirect gap of $E_{\text{gap}}^{I} = 0.68$ eV [4] (corresponding to $\lambda = 1.82\ \mu$m). However, for energies greater than 0.80 eV [5], direct transitions begin to occur in germanium. As a consequence of this transition change, the absorption coefficient (α) of germanium exhibits a significant increase, reflecting the greater probability of photon absorption. Figure 5.3 was constructed based on experimental data obtained by Nunley e*t al* [6], presenting the variation of the germanium absorption coefficient as a function of excitation energy. Note that α shows a discontinuity for $E_{\text{photon}} \approx E_{\text{gap}}^{\text{D}}$ and is null for $E_{\text{photon}} < E_{\text{gap}}^{I}$.

The absorption coefficient provides a measure of the distance that radiation needs to penetrate a material for its intensity to decay to $1/e$ of its incident value. In semiconductors, this absorption decay mechanism can occur through three main mechanisms [7]:

- α_{FC}—***through free carriers:*** This mechanism involves the absorption of radiation by free electrons in the conduction band and/or free holes in the valence band, leading to transitions to more energetic states (intraband

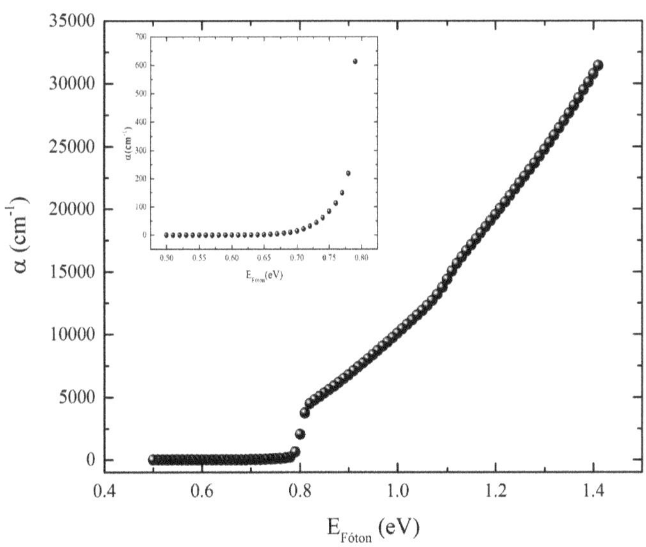

Figure 5.3. Main graph: Variation of the Germanium absorption coefficient as a function of excitation energy. Inset graph: Highlights the variation in the Germanium absorption coefficient for values of $E_{\text{photon}} < E_{\text{gap}}^{\text{D}}$. *Source*: The graphs were constructed based on experimental data obtained by Nunley e*t al* [6].

process). It dominates absorption processes when the carrier density is comparable to the electron density in metals.

- α_2—***through virtual processes with the absorption of two or more photons:*** This mechanism involves the absorption of two or more photons by an electron in the valence band, causing it to 'jump' to the conduction band and generating an electron–hole pair (interband process). It dominates absorption processes when the radiation intensity is high (as it is a non-linear process) and when the photon energy is insufficient for a vertical transition or lower than the energy of the material gap.
- α_1—***through the generation of charge carriers by the absorption of one photon band-to-band:*** This mechanism involves an electron in the valence band absorbing a photon and 'jumping' to the conduction band, generating an electron–hole pair (interband process). It dominates absorption processes outside of the previous two cases.

We can express the absorption coefficient α in terms of its components as

$$\alpha = \alpha_{FC} + \alpha_1 + \alpha_2. \tag{5.28}$$

Note that only α_1 and α_2 contribute to charge carrier generation.

We can estimate the density of photoinduced carriers using Beer's law. In this estimation, we assume that each absorbed (annihilated) photon generates an electron–hole pair. The induced carrier density decreases exponentially with distance from the material surface. The results estimated from Beer's law align with those obtained by Meyer e*t al* [7] through a more rigorous analysis, which involves the diffusion of carriers and the heating of the crystal lattice.

Beer's law is expressed as follows,

$$I(x) = I_0(x)e^{-ax}, \tag{5.29}$$

where $I(x)$ is the intensity of the radiation at depth x (measured from the surface), I_0 is the intensity of the incident beam, and α is the absorption coefficient of the material given by equation (5.28). As we aim to estimate the density of induced charge carriers, we can exclude absorption mechanisms that do not generate carriers, i.e. α_{FC}. Furthermore, considering that $E_{\text{photon}} \geqslant E^{D}_{\text{gap}}$, let us assume, for simplicity, that the dominant generation mechanism is due to α_1. Thus, we can approximate equation (5.28) by $\alpha = \alpha_1$.

Dividing both sides of equation (5.29) by $E_{\text{photon}} = \hbar\omega$ we have

$$\phi(x) = \phi_0 e^{-ax}, \tag{5.30}$$

where ϕ is the number of photons per unit area per unit time.

The rate of generation of carrier density $G(x)$ is therefore given by the variation of $\phi(x)$ when penetrating the material,

$$G(x) = -\frac{d\phi(x)}{dx} = \alpha\phi_0 e^{-\alpha(\lambda)x}. \tag{5.31}$$

The photon flux incident on the surface of the material for a Gaussian beam, as described by zero-order Gaussian field given by equation (2.78) (details in section 2.2.2), so we can write equation (5.30) as

$$\phi(0) = \phi_0 = \left(\frac{2P}{\pi w_0^2}\right)\frac{1}{\hbar\omega[1 + (2\tilde{z})^2]}\exp\left[-\frac{2(\tilde{x}^2 + \tilde{y}^2)}{1 + (2\tilde{z})^2}\right], \tag{5.32}$$

where $(\tilde{x}, \tilde{y}) = (x/w_0, y/w_0)$ are the normalized transverse coordinates and $\tilde{z} = z/kw_0^2$ is the normalized axial coordinate (direction of propagation). By replacing equation (5.32) in equation (5.31), we have the carrier density generation rate for a Gaussian beam,

$$G(x, y, z) = \left(\frac{2P}{\pi w_0^2}\right)\frac{\alpha}{\hbar\omega[1 + (2\tilde{z})^2]}\exp\left[-\frac{2(\tilde{x}^2 + \tilde{y}^2)}{1 + (2\tilde{z})^2}\right]. \tag{5.33}$$

The density of generated electrons (δn) and the density of generated holes (δp) is given by

$$\delta n = \delta p = G(x, \rho, z)\tau, \tag{5.34}$$

where τ is the lifetime (or recombination) time of the carriers. Using equation (5.33) into equation (5.34), we have

$$\delta n = \delta p = \left(\frac{2P}{\pi w_0^2}\right)\frac{\alpha\tau}{\hbar\omega[1 + (2\tilde{z})^2]}\exp\left[-\frac{2(\tilde{x}^2 + \tilde{y}^2)}{1 + (2\tilde{z})^2}\right]. \tag{5.35}$$

Thus, the carrier density optically excited in the semiconductor is given by

$$N = N_i + 2\delta n, \tag{5.36}$$

where N_i is intrinsic carrier density.

References

[1] Bochove E J and Walkup J F 1990 A communication on electrical charge relaxation in metals *Am. J. Phys.* **58** 131–4

[2] Fox M 2002 *Optical Properties of Solids* American Association of Physics Teachers

[3] Kittel C and McEuen P 2018 *Introduction to Solid State Physics* (New York: Wiley)

[4] Rezende S M 2004 *Materiais e dispositivos eletrônicos* Editora Livraria da Física

[5] Meyer J R, Kruer M R and Bartoli F J 1980 Optical heating in semiconductors: Laser damage in Ge, Si, InSb, and GaAs *J. Appl. Phys.* **51** 5513

[6] Nunley T N, Fernando N S, Samarasingha N, Moya J M, Nelson C M, Medina A A and Zollner S 2016 Optical constants of germanium and thermally grown germanium dioxide from 0.5 to 6.6 eV via a multisample ellipsometry investigation *J. Vac. Sci. Technol.* B **34** 061205

[7] Meyer J R, Bartoli F J and Kruer M R 1980 Optical heating in semiconductors *Phys. Rev.* B **21** 1559

IOP Publishing

Optical Trapping and Manipulation of New Materials

Tiago de Assis Moura, Joaquim Bonfim Santos Mendes and Márcio Santos Rocha

Chapter 6

Optical forces on semiconductors

Our goal in this chapter is to discuss the underlying physics involved in the optical manipulation of semiconductor particles using optical tweezers (OT). To achieve this, we will expand Ashkin's model [1] to account for the 'side effect' of carrier generation in semiconductor particles.

Carrier generation causes the polarizability of the particle to become a function of the beam intensity, leading to a spatial dependence that modifies the gradient force compared to what we obtained in chapters 2 and 4 for dielectric and plasmonic particles, respectively. The high extinction cross-section makes the optical traps stable in three dimensions only in the so-called plasmonic regime, which semiconductor particles reach when $\omega_p > \omega$.

For simplicity, we will focus on the Rayleigh regime, known for its analytical equations that are easily manipulated. The model developed in this chapter is applicable to any semiconductor particle in the Rayleigh regime, where the primary mechanism for carrier generation is through the band-to-band absorption of a single photon ($\alpha = \alpha_1$—see chapter 5).

6.1 Introduction

As discussed in chapter 5, the optics of semiconductors are significantly influenced by the density of charge carriers. This density can be altered through thermal excitation or by doping the semiconductor crystal lattice with elementary chemical impurities. Additionally, another method to modify the carrier density of a semiconductor is through optical absorption, which occurs when the semiconductor interacts with electromagnetic radiation.

Electron excitation takes place when the photon energy is sufficient to prompt the electron to 'jump' from a state in the valence band to a state in the conduction band. This transition can occur directly, where the electron does not change its wave vector (referred to as vertical transitions), or indirectly, where the electron changes its wave vector. In the latter case, due to the conservation of linear momentum, the process

doi:10.1088/978-0-7503-6074-6ch6

requires the absorption (or emission) of a phonon, in addition to the absorption of a photon. Consequently, if the photon energy exceeds the semiconductor energy gap, there is a probability that the photon will be absorbed, creating an electron–hole (e–h) pair and increasing the carrier density of the semiconductor.

Unlike insulators and conductors, semiconductor materials exhibit considerable variability in their carrier density. Therefore, when exposed to the laser beam of an OT (here, we will consider the case where $\lambda = 1064$ nm, $E_{\text{photon}} = 1.15$ eV), semiconductor materials are expected to undergo modifications in their carrier densities due to the absorption of laser photons.

As demonstrated in section 5.3.1, there are three primary optical absorption mechanisms in semiconductors (α_1, α_2, and α_{FC}), and the dominant mechanism depends on factors such as the band structure of the semiconductor, the presence of impurities in the crystal lattice, and radiation properties like wavelength and intensity. Semiconductors with a direct gap smaller than the photon energy, such as germanium (with $E_{\text{gap}} = 0.80$ eV at 300 K in relation to $\lambda = 1064$ nm) [2], will have α_1 as the dominant absorption mechanism. In contrast, semiconductors with an indirect gap, like silicon ($E_{\text{gap}} = 1.11$ eV) [3], or with a direct gap greater than the photon energy, such as gallium arsenide ($E_{\text{gap}} = 1.42$ eV) [4], will exhibit virtual absorption processes α_2 as their dominant absorption mechanism at $\lambda = 1064$ nm, especially in strongly focused beams like those used in OT [5].

Note that the laser in OT serves a dual role in semiconductors particles: it acts as an element of the optical trap, attracting the particle to the focal region[1] through the gradient force, and simultaneously acts as an exciting agent for the material, inducing the generation of charge carriers, as illustrated in figure 6.1.

In addition to modifying the optics of semiconductors, carrier generation can elevate the temperature of the semiconductor crystal lattice through the thermalization of hot carriers (induced or due to the absorption of free carriers) and, eventually, non-radiative recombination of e–h pairs. If $E_{\text{photon}} > E_{\text{gap}}$, a fraction of the absorbed energy will always be converted into heat, regardless of the type of recombination the carriers undergo. Thus, thermal effects will inevitably occur when manipulating semiconductors using OT.

The key question is: does this heating generate a temperature gradient on the particle surface sufficient to produce a photophoretic force capable of overcoming the gradient force and inducing the observed and well characterized oscillatory motion that such particles usually present in OT—see chapter 7 and references [4, 6]? Such a scenario was proposed when observing the phenomenon with Bi_2Te_3 and Bi_2Se_3 particles [4], where the gap energy is significantly lower than the laser photon energy ($E_{\text{gap}} = 0.22$ eV) [7], and the absorption coefficient is comparable to that of metals ($\alpha = 1.4 \times 10^5$ eV) [8].

[1] We will see later that the gradient force does not exactly drive the particle toward the focal region but rather to the location where the optical potential is minimized. However, considering that the initial carrier density is small, we can approximate the region where the optical potential is minimized to the region where the field intensity is maximum, that is, the focal region.

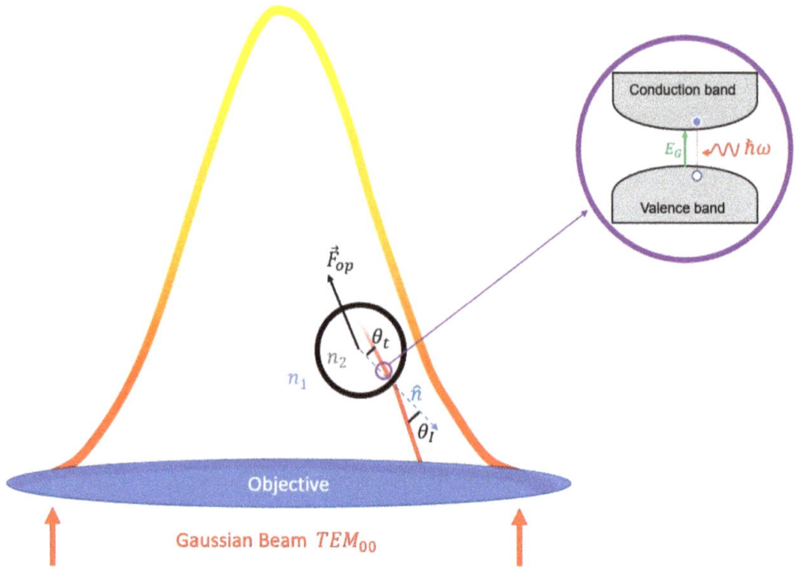

Figure 6.1. Scheme illustrating the dual role played by the laser: constructing the optical trap that guides semiconductor particles to the focal region and acting as an exciting agent for the material, inducing the generation of charge carriers.

However, as the phenomenon was observed in semiconductors with bandgap energies closer to the photon energy and lower absorption coefficients, such as germanium and especially silicon, our attention turned to assessing the effect of changes in the optical properties of semiconductors due to carrier generation. Specifically, in reference [9] we proposed that the gradient force becomes repulsive beyond a certain position to explain the oscillations observed in silicon (see also the discussion in chapter 7). The remainder of this chapter aims to demonstrate more robustly how the gradient force becomes repulsive, the conditions necessary for this to occur, and the impact on extinction forces. In other words, we will evaluate the effect of carrier generation on optical forces.

6.2 The model

Let us start by considering the interaction of a spherical semiconductor particle (immersed in water, $\varepsilon_m = 1.77$) with a zeroth-order Gaussian beam described by $w_0 = 5\ \mu$m and $\lambda = 1064$ nm. According to Davis [10] and Barton [11, 12], the zeroth-order description presents average errors of 1% when compared to higher-order descriptions of the Gaussian beam ($s = 0.025$, see details in section 2.2.2).

We will adopt a Cartesian coordinate system with the origin at the beam waist, as used in chapter 2 (see figure 2.9). The intensity profile of the beam as a function of the normalized coordinates $(\tilde{x}, \tilde{y}, \tilde{z}) = (x/w_0, y/w_0, z/kw_0^2)$ is shown in figure 6.2.

Our intention in this chapter is to evaluate how the generation of charge carriers alters the optical forces acting on a semiconductor particle. To accomplish this, we

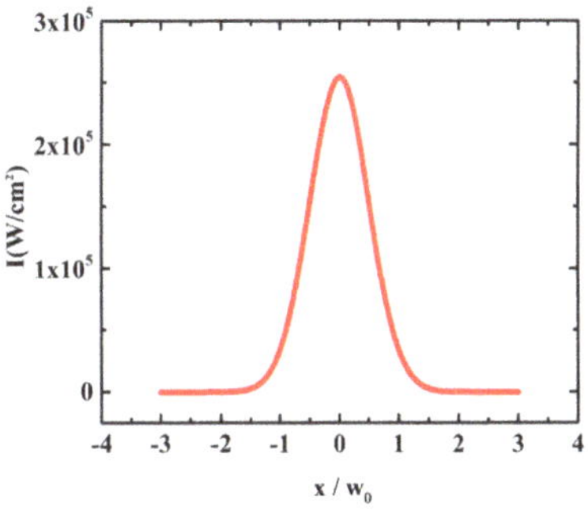

(a) Transverse intensity profile.

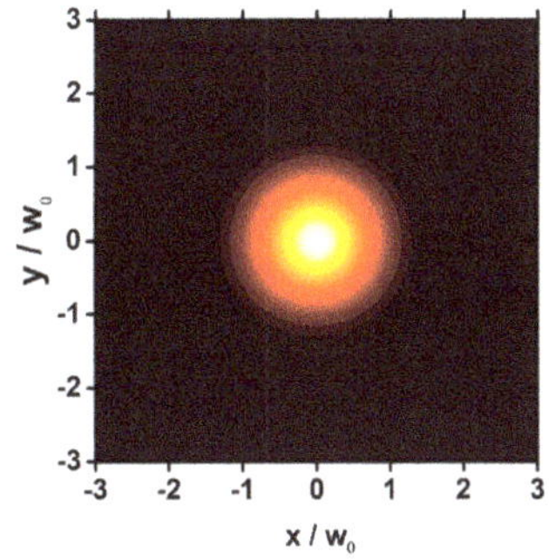

(b) Cross-sectional intensity profile.

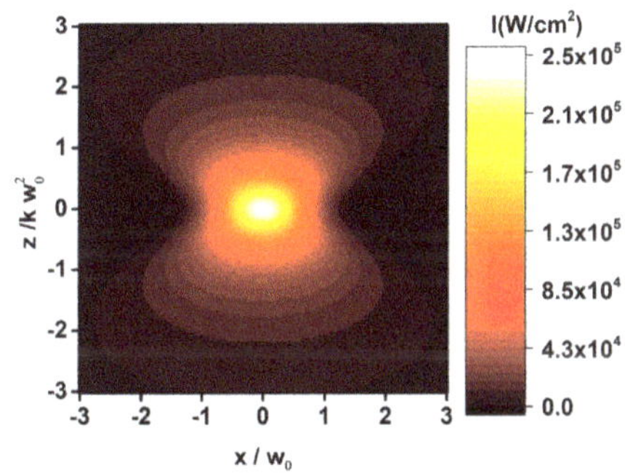

(c) Longitudinal intensity profile in the zx plane

Figure 6.2. Zero-order Gaussian field intensity profile with $w_0 = 5\ \mu$m. (a) Variation of intensity as a function of the normalized transverse position x/w_0. $y = 0$, $z = 2.5\ \mu$m. (b) Transversal intensity profile, $z = 2.5\ \mu$m. (c) Variation of the intensity profile depending on the normalized positions x/w_0 and z/kw_0^2, $y = 0$. $w_0 = 5\ \mu$m, $P = 100$ mW.

will use the concepts presented in section 5.3. Carrier generation is primarily governed by Beer's absorption law, where each absorbed photon generates an e–h pair, and the injected carrier density decreases exponentially with the distance from the material surface (ξ).

Considering that the absorption coefficient is $\alpha = \alpha_1 = 1.9 \times 10^4\ \text{cm}^{-1}$ (a typical value for a direct bandgap semiconductor like germanium) [13], we can calculate the generation rate (G) from the carrier density at the surface of the semiconductor particle using the following equation (5.33),

$$G(x, y, z, \xi) = -\frac{1}{\hbar\omega}\frac{d[I(x, y, ze^{-\alpha_1\xi})]}{d\xi} = \frac{\alpha_1 I(x, y, z)e^{-\alpha_1\xi}}{\hbar\omega}. \tag{6.1}$$

Note that for particles whose diameter is much smaller than the penetration length ($\nu = \alpha^{-1} \simeq 500\quad \text{nm}$), we can approximate the generation rate in the bulk as equal to the generation rate at the surface ($\xi = 0$). Therefore, we can simplify equation (6.1) to

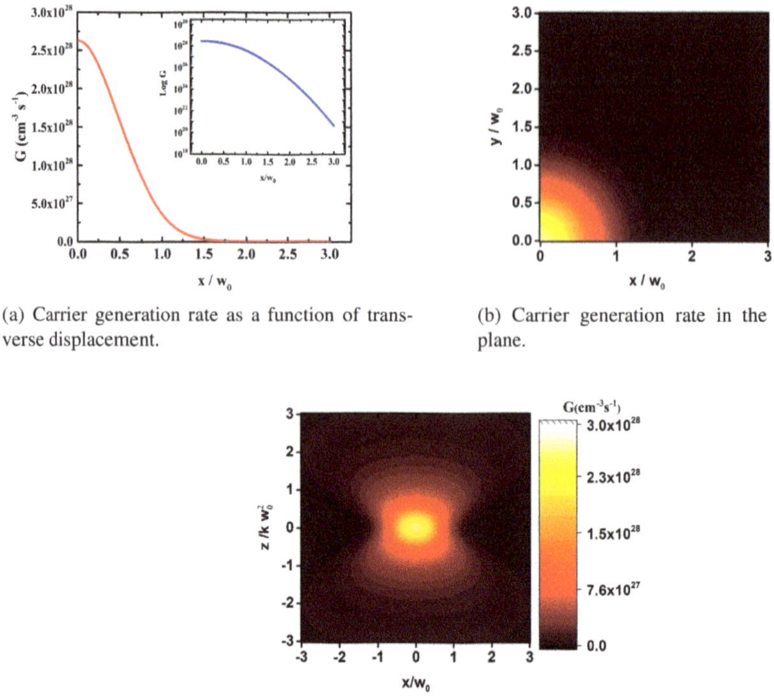

(a) Carrier generation rate as a function of transverse displacement.

(b) Carrier generation rate in the transverse xy plane.

(c) Carrier generation rate in the xz plane

Figure 6.3. Carrier generation rate in the semiconductor particle illuminated by the OT. (a) Variation of rate generation as a function of the normalized transverse position x/w_0. $y = 0$, $z = 2.5\ \mu$m. *Inserted graph:* Variation of log G as a function of x/w_0. (b) Rate generation induced on the semiconductor particle as a function of its normalized transverse position under OT illumination, $z = 2.5\ \mu$m. (c) Rate generation as a function of its the normalized position x/w_0 and z/kw_0^2, $y = 0$. $w_0 = 5\ \mu$m, $P = 100$ mW, $\hbar\omega = 1.15$ eV, $\alpha = 1.9 \times 10^4\ \text{cm}^{-1}$.

$$G(x, y, z) = \left(\frac{2P}{\pi w_0^2}\right)\frac{\alpha}{\hbar\omega[1 + (2\tilde{z})^2]}\exp\left[-\frac{2(\tilde{x}^2 + \tilde{y}^2)}{1 + (2\tilde{z})^2}\right], \tag{6.2}$$

where $\hbar\omega = 1.15$ eV is the photon energy for $\lambda = 1064$ nm. Figure 6.3 illustrates the carrier generation rate of a semiconductor particle under the illumination of a zeroth-order Gaussian beam.

To obtain the carrier density generated by optical absorption, we must multiply the generation rate given by equation (6.2) by the carrier recombination time (τ) as follows

$$\delta e = \delta h = \delta n = G(x, y, z)\tau, \tag{6.3}$$

where δe and δh are the densities of electrons and holes created, respectively. The total carrier density as a function of the particle position can be obtained by adding the intrinsic carrier density of the semiconductor ($N_i = 10^{13}\ \text{cm}^{-3}$ for germanium [2]) to equation (6.2),

$$N = N_i + 2\delta n, \tag{6.4}$$

$$N(x, y, z, \tau) = N_i + \left(\frac{4P}{\pi w_0^2}\right)\frac{\alpha\tau}{\hbar\omega[1 + (2\tilde{z})^2]}\exp\left[-\frac{2(\tilde{x}^2 + \tilde{y}^2)}{1 + (2\tilde{z})^2}\right]. \tag{6.5}$$

As discussed in chapter 5, the type of recombination is crucial in determining how the absorbed energy is converted. Additionally, N is proportional to the carrier lifetime (another way of referring to the recombination time). Therefore, the longer the lifetime, the higher the carrier density for the same generation rate.

A characteristic of radiative recombination is its long lifetimes, typically greater than μs, whereas carriers that recombine non-radiatively (mostly through Auger recombination) have lifetimes on the order of nanoseconds. The type of recombination (and thus the carrier lifetime) depends on several factors such as the semiconductor band structure [14], radiation intensity [5], carrier density [15], and the surrounding medium [16], among others.

For germanium, recombination times on the order of tens of nanoseconds have been measured. Driel and Galante [15] reported a recombination time of approximately 50 ns, while Tan *et al* [17] measured a recombination time of 70.9 ns for a carrier density of 2.4×10^{21} cm^{-3}. These values are associated with non-radiative recombination, specifically Auger recombination, indicating that a significant portion of the energy absorbed by germanium will be converted into heat.

Figure 6.4 illustrates the carrier generation within the semiconductor particle for various carrier lifetime values.

Note that the carrier density generated by photoexcitation is significantly higher than the intrinsic carrier density of the semiconductor ($N_i = 10^{13}$ cm^{-3}). In fact, these values are consistent with those reported in the literature. Galante and Driel [15] obtained a carrier density of $N = 10^{19}$ cm^{-3} when illuminating a germanium wafer ($d = 400$ μm thick) with a pulsed laser ($\tau_p = 80$ ns, $\lambda = 1064$ nm, $I = 2 \times 10^5$ W cm^{-2}), measuring the variation in reflectivity of a probe laser ($\lambda = 10.6$ μm) on the wafer during the source laser pulse.

The technique used limits the detection of carrier density to approximately $N = 10^{19}$ cm^{-3} due to resonance between the plasma frequency and the probe laser frequency [18]. Carrier densities of around $N = 3.5 \times 10^{20}$ cm^{-3} were obtained by Yeh *et al* [19] using a probe laser at $\lambda = 1.55$ μm in a setup similar to that of Galante and Driel. Once again, the measured density is close to the plasma resonance of the probe ($N_R = 6.8 \times 10^{20}$ cm^{-3}).

In this chapter, we will use the carrier lifetime (τ) as a parameter to determine the carrier density generated in the semiconductor particle. We make this choice because τ exclusively affects the carrier density without altering the characteristics of the optical trap. This simplifies the analysis of the impact of carrier generation on optical forces, allowing this effect to be isolated.

From an experimental perspective, the most direct control of the carrier density is achieved by varying the beam power (P). However, changing P also modifies the trap intensity, making it more difficult to separate the effects. Since the carrier

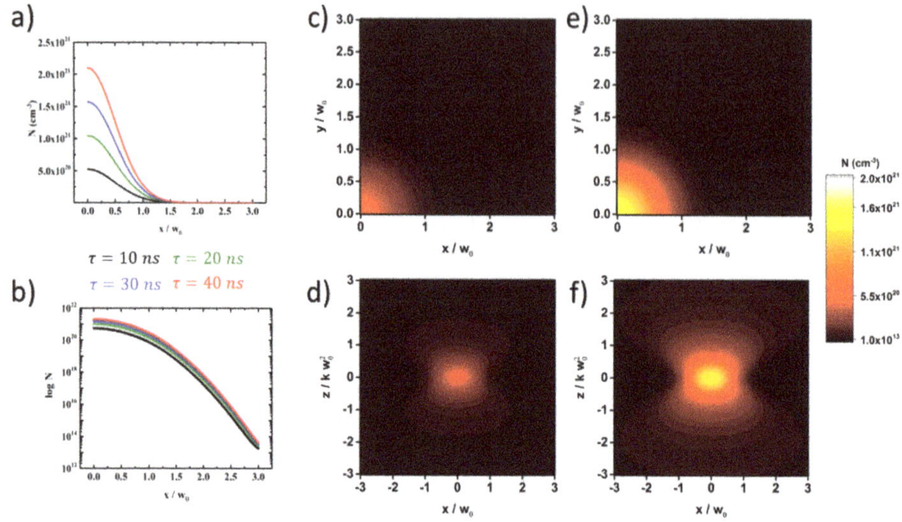

Figure 6.4. Resultant carrier density induced on the semiconductor particle illuminated by the OT. (a and b) Variation of carrier density as a function of the normalized transverse position x/w_0. $y = 0$, $z = 2.5\ \mu$m, for various lifetime values. (c) Resulting carrier density induced on the semiconductor particle as a function of its normalized transverse position under OT illumination for a recombination time of $\tau = 10$ ns. $z = 2.5\ \mu$m. (d) Resultant carrier density as a function of its the normalized position x/w_0 and z/kw_0^2, $y = 0$ for $\tau = 10$ ns. (e) Same as (c) for $\tau = 30$ ns. (f) Same as d for $\tau = 30$ ns. $w_0 = 5\ \mu$m, $P = 100$ mW, $\hbar\omega = 1.15$ eV, $\alpha = 1.9 \times 10^4$ cm^{-1}, $N_i = 10^{13}$ cm^{-3}.

lifetime is not an adjustable parameter in an experiment, the reader should interpret it here as a theoretical variable representing different induced carrier densities, without directly influencing the optical trap.

Note that the carrier density in the semiconductor particle increases by approximately eight orders of magnitude as it approaches the region of highest beam intensity. This significant increase directly impacts the optical properties of the material.

To describe this influence, we will use the Drude model, which allows us to incorporate the effect of free carriers into the effective dielectric constant of the semiconductor (ε_r), as discussed in section 5.3.

We consider that, in the absence of illumination, the optical properties of the semiconductor are described by $\varepsilon_{opt} = \varepsilon_{opt}' + i\varepsilon_{opt}''$, representing the response of bound electrons (P_{bound}). Since the effect of intrinsic free carriers on ε_r is negligible, all modifications caused by the generation of optically excited carriers will be attributed to the contribution of free carriers (P_{free}) in equation (5.18).

The variation in carrier density directly modifies the plasma frequency of the semiconductor (ω_p), which can be expressed by equation (6.6):

$$\omega_p = \sqrt{\frac{Ne^2}{\varepsilon_{opt}'\varepsilon_0 m^*}}, \tag{6.6}$$

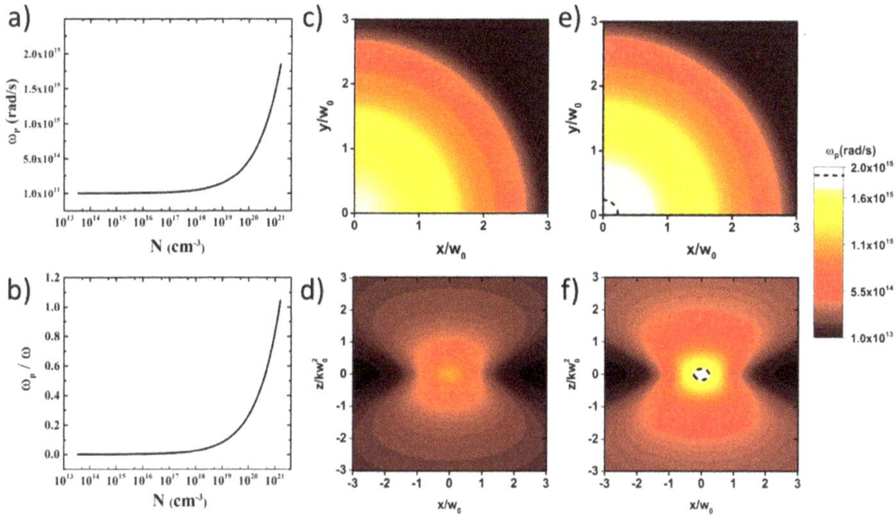

Figure 6.5. Plasma frequency induced on the semiconductor particle illuminated by the OT. (a) Variation of plasma frequency as a function of carrier density. (b) Plasma frequency normalized by the OT laser frequency as a function of carrier density. (c) Plasma frequency induced on the semiconductor particle as a function of its normalized transverse position under OT illumination for a recombination time of $\tau = 10$ ns. $z = 2.5\ \mu$m. (d) Plasma frequency induced in the semiconductor particle as a function and its normalized position z/w_0 and z/kw_0^2 under OT illumination for a recombination time of $\tau = 10$ ns. $y = 0$. (e) Same as (c) for $\tau = 30$ ns. (f) Same as (d) for $\tau = 30$ ns. The dashed line indicates the positions where $\omega_p = \omega = 1.75 \times 10^{15}$ rad s^{-1}.

where e is the elementary charge, m^* is the reduced effective mass of the carriers (for germanium, $m^*=0.076m_e$, with m_e being the electron mass), and N is the carrier density.

Since the laser frequency is fixed, increasing the carrier density N raises ω_p, thus changing the ratio ω_p/ω. Figures 6.5(a) and (b) illustrate, respectively, the plasma frequency and the ω_p/ω ratio as functions of the carrier density.

It is important to highlight that plasma resonance ($\omega_p = \omega_{1064}$) is reached for $N = 1.41 \times 10^{21}$ cm^{-3}. This carrier density can be achieved by illuminating the semiconductor particle with a zeroth-order Gaussian beam, as previously shown in figure 6.4.

Figures 6.5(c) and (d) present the spatial variation of the plasma frequency along the normalized position of the semiconductor particle for a carrier lifetime $\tau = 10$ ns. Similarly, figures 6.5(e) and (f) show the same behavior for $\tau = 30$ ns.

Interestingly, for $\tau = 30$ ns, plasma resonance is reached near the focal region (indicated by the dashed line). While in typical plasmonic systems the incident light frequency is tuned to match the material's plasma frequency, here the situation is reversed: the laser frequency is fixed, and it is the plasma frequency of the semiconductor that becomes tunable through the optically generated carrier density. This tunability can be controlled by illumination conditions, material doping, or other techniques.

Through the Drude model, we can analyze how the variation in the plasma frequency of the semiconductor influences its optical properties, particularly its dielectric constant, which can be expressed by equation (6.7),

$$\varepsilon_{\mathrm{r}}(\omega) = \varepsilon_{\mathrm{opt}}'\left(1 - \frac{\omega_{\mathrm{p}}^2}{\omega^2 + i\gamma\omega}\right) + i\varepsilon_{\mathrm{opt}}''. \tag{6.7}$$

Figure 6.6 illustrates how the dielectric constant of the semiconductor particle varies as a function of carrier density, a variation resulting from the illumination of the particle by the Gaussian beam, as shown in figure 6.4.

Note that the primary effect on $\mathrm{Re}\{\varepsilon_{\mathrm{r}}\}$ (figure 6.6(a)) is its reduction, becoming zero at the plasma resonance condition ($\omega_{\mathrm{p}} = \omega$), indicated by the green dashed line. When $\omega_{\mathrm{p}} > \omega$, we have $\mathrm{Re}\{\varepsilon_{\mathrm{r}}\} < 0$, and the semiconductor starts behaving as a single-negative medium (SNG), governed by the principles discussed in chapter 4.

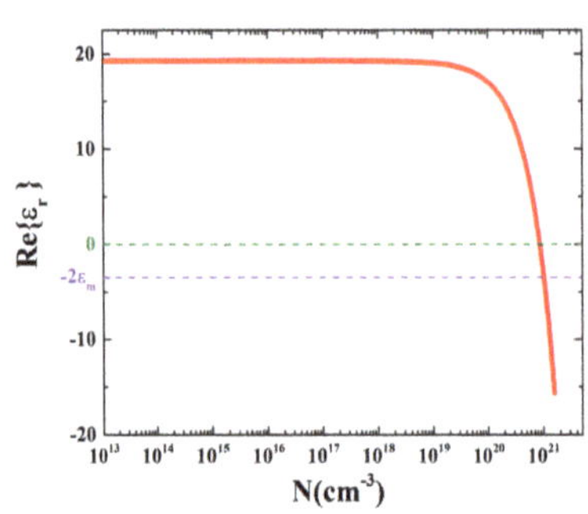

(a) $\mathrm{Re}\{\varepsilon_r\}$ as a function of N

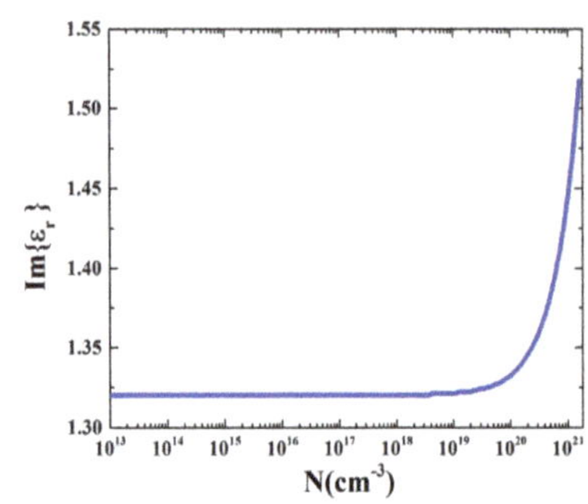

(b) $\mathrm{Im}\{\varepsilon_r\}$ as a function of N

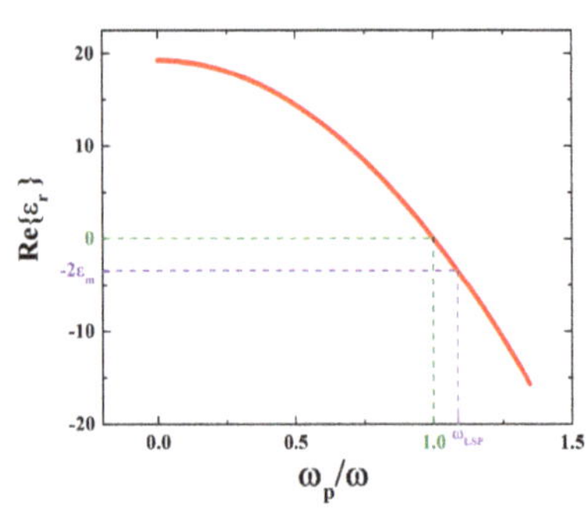

(c) $\mathrm{Re}\{\varepsilon_r\}$ as a function of ω_p/ω

Figure 6.6. Variation in the dielectric constant of a semiconductor as a function of carrier density. (a) Real part. (b) Imaginary part. (c) Variation in the dielectric constant of a semiconductor as a function of normalized plasma frequency. The green dashed line indicates plasma resonance, while the purple dashed line indicates LSP resonance. The intrinsic dielectric constant of the material at $\lambda = 1064$ nm is $\varepsilon_{\mathrm{opt}} = 19.2 + 1.32i$, with $\omega = 1.77 \times 10^{15}$ rad s^{-1}.

In this regime, free carriers on the particle surface can couple with the electric field of light, behaving as localized surface plasmons (LSPs). The LSP resonance is achieved when the Fröhlich condition is satisfied, that is, when $\mathrm{Re}\{\varepsilon_r\} = -2\varepsilon_m$, represented by the purple dashed line.

Using equation (6.7), the Fröhlich condition is satisfied when[2]

$$\omega_p = \sqrt{\frac{\varepsilon_{opt}{}' + 2\varepsilon_m}{\varepsilon_{opt}{}'}}\,\omega. \tag{6.8}$$

For germanium particles immersed in water, the LSP resonance is achieved when $\omega_p \approx 1.09\ \omega_{1064}$, as illustrated in figure 6.6(c).

This variation in the dielectric constant directly impacts other optical properties of the semiconductor, such as the refractive index, reflectivity, and absorption coefficient, as illustrated in figure 5.2. It is worth noting that the increase in the absorption coefficient is due to free-carrier absorption, which does not contribute to the generation of new carriers. Therefore, there is no additional increase in the carrier generation rate associated with this process.

Before we understand why free-carrier generation decreases the real part of the dielectric constant, let us analyze the polarizability of the semiconductor particle and how it is modified by carrier generation. In chapter 2, we discussed the polarization process in a dielectric particle within the quasi-static approximation, and in chapter 4, we extended this understanding to plasmonic particles. Here, we will present the particularities that arise when the particle is a semiconductor. All the necessary mathematics have already been developed in previous chapters; we will not start from scratch but rather interpret the results considering the material as a semiconductor. Our focus now is to understand how the presence of free carriers alters this classical polarization response.

Consider the situation illustrated in figure 2.9, where we now have a semiconductor particle of radius a, much smaller than the wavelength λ, interacting with a linearly polarized zeroth-order Gaussian beam (along the x-axis) propagating in the z direction. Initially, we assume that the field intensity is very low or that the carrier lifetime is very short, so that carrier generation is negligible and the effect of the field acts only on the bound electrons. When interacting with the electric field, the electronic cloud (negatively charged) shifts relative to the positive nucleus, generating a dipole moment density given by equation (2.9) and illustrated in figure 2.2(b).

In this configuration, there is no significant difference between a semiconductor particle and a dielectric particle, so we can use the set of equations from chapter 2 to

[2] It is worth noting that equation (6.8) provides the plasma frequency required for LSP resonance to occur exactly at the incident light frequency ω. Since, in this context, we consider the light frequency fixed and adjust the plasma frequency through carrier generation, throughout this chapter, when referring to ω_{LSP}, we will be indicating the value that ω_p must assume for LSP resonance to occur at ω.

analyze the problem, simply replacing $\varepsilon_p = \varepsilon_{opt} = \varepsilon_r$. In particular, the induced dipole moment is given by equation (2.22), whose polarizability can be written as[3]

$$\Gamma = 4\pi a^3 \frac{\varepsilon_r - \varepsilon_m}{\varepsilon_r + 2\varepsilon_m}. \tag{6.9}$$

Just like in dielectrics, these dipoles orient themselves inside the particle in such a way that they cancel out, leaving a charge on the surface, whose density can be expressed as

$$\sigma_{pol} = 3\varepsilon_0 \left(\frac{\varepsilon_r - \varepsilon_m}{\varepsilon_r + 2\varepsilon_m} \right) E_0 \cos\theta, \tag{6.10}$$

where E_0 is the amplitude and θ is the angle between the polarization direction of the light's electric field and the position vector $\mathbf{r}$ on the particle surface, with $\theta = 0$ referring to the particle north pole and $\theta = \pi$ to the south pole[4].

For a typical (undoped) semiconductor, such as germanium or silicon, we have $\mathrm{Re}\{\varepsilon_r\} > \varepsilon_m$ (with $\varepsilon_m = 1.77$). Thus, the induced charges are arranged so that the north pole is positively charged, making the induced dipole moment parallel to the electric field of light, just as in a dielectric particle with $\varepsilon_p > \varepsilon_m$.

However, if $E_{photon} \geqslant E_{gap}$, the semiconductor nature manifests itself, and as the beam intensity increases—e.g., when the particle approaches the optical axis of a Gaussian beam—the number of free charge carriers increases, reducing $\mathrm{Re}\{\varepsilon_r\}$ and, consequently, Γ.

Figure 6.7(a) shows how the real part of the polarizability of a semiconductor particle ($a = 15$ nm) varies as a function of transverse displacement when illuminated by a Gaussian beam with $P = 100$ mW. For short lifetimes ($\tau \leqslant 25$ ns), as the particle approaches the optical axis, $\mathrm{Re}\{\Gamma\}$ decreases until it approaches zero but remains positive. The higher the value of τ within this range, the lower the value of $\mathrm{Re}\{\Gamma\}$.

For $\tau = 30$ ns, as the particle approaches the optical axis, in addition to decreasing, $\mathrm{Re}\{\Gamma\}$ becomes negative, indicating an inversion in the way charges are induced on the particle surface. For $\tau = 35$ ns, a non-monotonic behavior is observed: as it approaches the optical axis, $\mathrm{Re}\{\Gamma\}$ first decreases, reaches negative values, and then abruptly increases, becoming positive again.

Analyzing the imaginary part of Γ (figure 6.7(b)), we observe an increasing behavior as the particle approaches the optical axis for $\tau \leqslant 30$ ns and a peak off the optical axis for $\tau = 35$ ns.

To interpret this result, let us analyze what is happening with the charge density on the surface of the semiconductor particle. Figure 6.7(c) shows the induced charge

[3] Note that, for infrared frequencies, the radiation correction in polarizability (equation (2.42)) is negligible for particles in the Rayleigh condition, since $ka \ll 1$.

[4] To aid in visualizing the problem, the reader may revisit figure 2.1(b), considering only that, in the current scenario, the polarization direction of the electric field is along the x-axis rather than the z-axis, as illustrated in the figure.

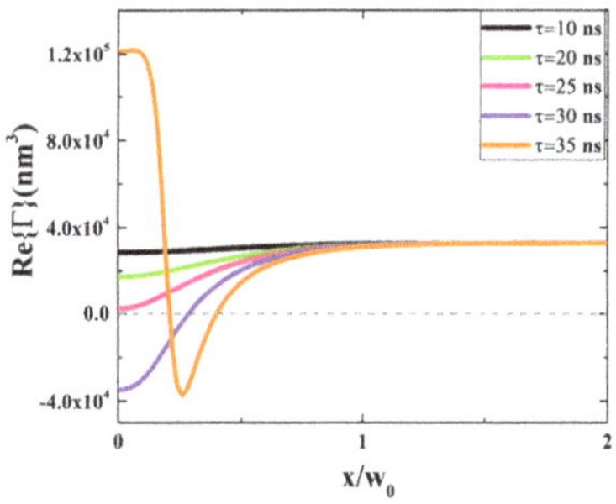

(a) Re{Γ} as a function of normalized transverse displacement.

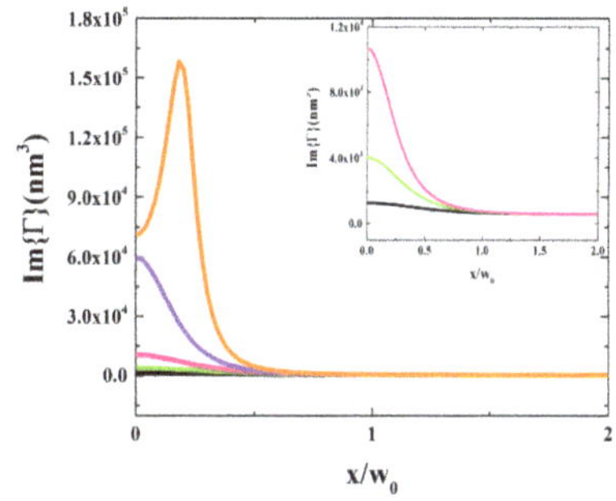

(b) Im{Γ} as a function of normalized transverse displacement.

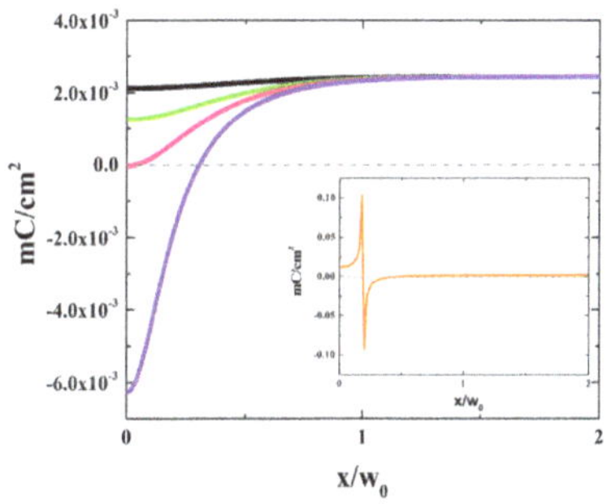

(c) Induced surface charge density at $\theta = 0$.

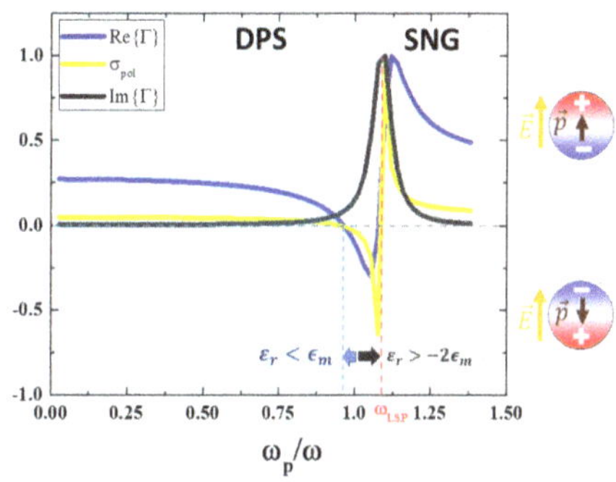

(d) Normalized Re{Γ}, Im{Γ}, σ_{pol} as a function ω_p/ω.

Figure 6.7. Effect of carrier generation on the polarizability and induced charge density in the semiconductor particle. (a) Real part of the polarizability as a function of the normalized position along the x-axis for different carrier lifetimes. (b) Imaginary part of the polarizability. (c) Surface charge density induced at the north pole ($\theta = 0$) of the semiconductor particle for different carrier lifetimes. Inset: induced charge density at the LSP resonance. (d) Polarizability (real and imaginary parts) and charge density (north pole) normalized as a function of the plasma frequency normalized by ω. The blue region represents when the particle behaves as a DPS medium, while the gray region represents when the particle behaves as an SNG medium. Positive values of Re{Γ} and σ_{pol} indicate that the induced dipole moment is parallel to the beam's electric field, whereas negative values indicate that the dipole moment is antiparallel to the electric field. *Parameters:* $a = 15$ nm; $P = 100$ mW; $\lambda = 1064$ nm; $w_0 = 5\ \mu$m; $\mathbf{r}_\text{d} = (x/w_0, 0, 0)$. The semiconductor properties are the same as those indicated in figure 6.4.

density at the north pole of the particle ($\theta = 0$). Notice that, as the particle approaches the optical axis, σ_{pol} decreases, and the larger τ is, the lower the surface charge density. As anticipated, for $\tau = 30$ ns, the charge density at the north pole becomes negative near the optical axis.

But why does carrier generation decrease the surface charge density? If we are increasing the density of free charges, shouldn't the opposite happen?

The answer is no! Let us see why. When the carrier density increases, these carriers begin to shield the particle from the external electric field through an effect known as *screening* or *shielding*[5]. This behavior is characteristic of semiconductor materials, in which the free-carrier density can be modulated by illumination, allowing direct control over the particle optical response. In other words, as the carrier density grows, the carriers distribute themselves throughout the particle in such a way as to reduce the internal electric field. Consequently, the displacement between the electron cloud and the nucleus decreases, which is reflected in the reduction of surface charge density.

When $\varepsilon_r = \varepsilon_m$, the permittivity contrast disappears, and the particle 'camouflages' itself in the medium—the field cannot polarize it. As a result, $\sigma_{pol} = 0$ and $\mathrm{Re}\{\Gamma\} = 0$. If more carriers are induced, we get $\varepsilon_r < \varepsilon_m$, so that the dipole moment inverts (becoming antiparallel to the electric field), and negative charges are induced at the north pole, leading to the observed negative values.

When $\omega_p = \omega$, $\mathrm{Re}\{\varepsilon_r\} = 0$, and the semiconductor transitions from being a DPS (double-positive medium) to an SNG medium. As an SNG medium, the semiconductor can no longer support the propagation of electromagnetic waves, and the surface charge density becomes governed by the free carriers, just as the surface charge density in a plasmonic particle is defined by free electrons, as discussed in chapter 4. This process highlights how free-carrier generation not only alters the surface charge but also modifies the entire optical response of the particle, directly influencing its polarizability and effective dielectric constant.

Note that the surface charge density is negative during the DPS phase due to bound electrons, since $\varepsilon_r < \varepsilon_m$. Now, the surface charge density is negative in the SNG phase because $\varepsilon_r > -2\varepsilon_m$. In these scenarios, the charges are oriented such that the induced dipole moment opposes the electric field, as observed in figures 2.4 (dielectrics) and 4.9 (metal).

The behavior of $\mathrm{Im}\{\Gamma\}$ can be easily understood from figure 5.2(d). As the density of free carriers increases, the absorption contribution of free carriers, represented by α_{FC}, also grows, resulting in an increase in optical losses. Consequently, $\mathrm{Im}\{\Gamma\}$ progressively rises until it reaches a peak at the LSP resonance, where losses are maximized due to strong charge accumulation and efficient coupling with the external electric field.

Figure 6.7(d) summarizes our entire discussion by illustrating the variation of the normalized values of $\mathrm{Re}\{\Gamma\}$, $\mathrm{Im}\{\Gamma\}$, and the surface charge density σ_{pol} as a function of the plasma frequency normalized by the incident beam frequency, ω_p/ω. The transition between the DPS (blue) and SNG (gray) regimes occurs when $\omega_p = \omega$.

In the DPS regime, the particle behaves like a dielectric with a positive permittivity higher than that of the surrounding medium when $\varepsilon_r > \varepsilon_m$. That is, the light electric field polarizes the particle, aligning the induced dipole moment with the applied field. As ω_p increases and ε_r approaches ε_m, the permittivity contrast

[5] The screening effect (or shielding effect) occurs when free charges within a material rearrange themselves to reduce the applied electric field inside it. In semiconductors, as the density of free carriers increases, these carriers quickly respond to the external field, creating an opposing internal field that 'shields' the applied field. As a result, the effective field experienced by bound carriers decreases, reducing the displacement of the electron cloud and, consequently, the polarization of the particle.

decreases, leading σ_{pol} and $\mathrm{Re}\{\Gamma\}$ to values close to zero. When the condition $\varepsilon_r = \varepsilon_m$ is met, the particle becomes practically invisible to the external field, resulting in $\sigma_{pol} = 0$ and $\mathrm{Re}\{\Gamma\} = 0$, while $\mathrm{Im}\{\Gamma\}$ remains positive due to absorption (losses) by free carriers.

With a further increase in ω_p, the condition $\varepsilon_r < \varepsilon_m$ is satisfied, inverting the sign of the induced dipole moment, which becomes antiparallel to the applied electric field. Within the DPS regime, the particle dipole moment is primarily due to bound electrons, with free carriers playing the role of shielding the external field.

Upon reaching the SNG regime, defined by the condition $\omega_p > \omega$, the dipole moment—and thus the entire polarization process—is dominated by free carriers. For values of ω_p where $\varepsilon_r > -2\varepsilon_m$, the free carriers distribute themselves on the particle surface in such a way as to generate an induced dipole moment antiparallel to the light electric field, resulting in negative polarizability (real part) and negative surface charge density (at the north pole of the particle). When the resonance condition $\varepsilon_r = -2\varepsilon_m$ is met, the LSP occurs, with maximum surface charge accumulation and the polarizability reaching its peak—the real part reflecting the strong charge induction and the imaginary part marking the peak in optical losses. For $\omega_p > \omega_{LSP}$, the free carriers reorganize on the particle surface, re-establishing the alignment of the induced dipole moment with the incident light electric field.

The origin of $\mathrm{Im}\{\Gamma\}$ lies in the extinction of light by the particle, which involves both the absorption and scattering of the incident radiation. In chapter 2, we defined the extinction cross-section (Q_{ext}) as the ratio between the extinguished power and the intensity of the incident wave. This quantity has units of area and represents the effective area with which the particle interacts with light. The extinction, and consequently the extinction cross-section, accounts for all the losses occurring during the interaction of the particle with light, with absorption and scattering being the primary ones.

Following the same procedure adopted in chapters 2 and 4, we can define the absorption and scattering cross-sections as the ratio between the absorbed (or scattered) power and the intensity of the incident beam. Using equations (2.49) and (2.57), we obtain

$$Q_{ext} = k\,\mathrm{Im}\{\Gamma\}, \tag{6.11}$$

$$Q_{scat} = \frac{k^4}{6\pi}|\Gamma|^2, \tag{6.12}$$

$$Q_{abs} = Q_{ext} - Q_{scat}, \tag{6.13}$$

where k is the wave number in the medium.

Figure 6.8(a) shows the cross-sections of a semiconductor particle with a radius of 15 nm for $\tau = 10$ ns. In this regime, the carrier density is not sufficient to reach the LSP resonance, and the particle behaves as a dielectric. It is observed that the extinction cross-section (solid red line) is dominated by the absorption cross-section (dashed green line), which is expected, since it is through absorption that charge

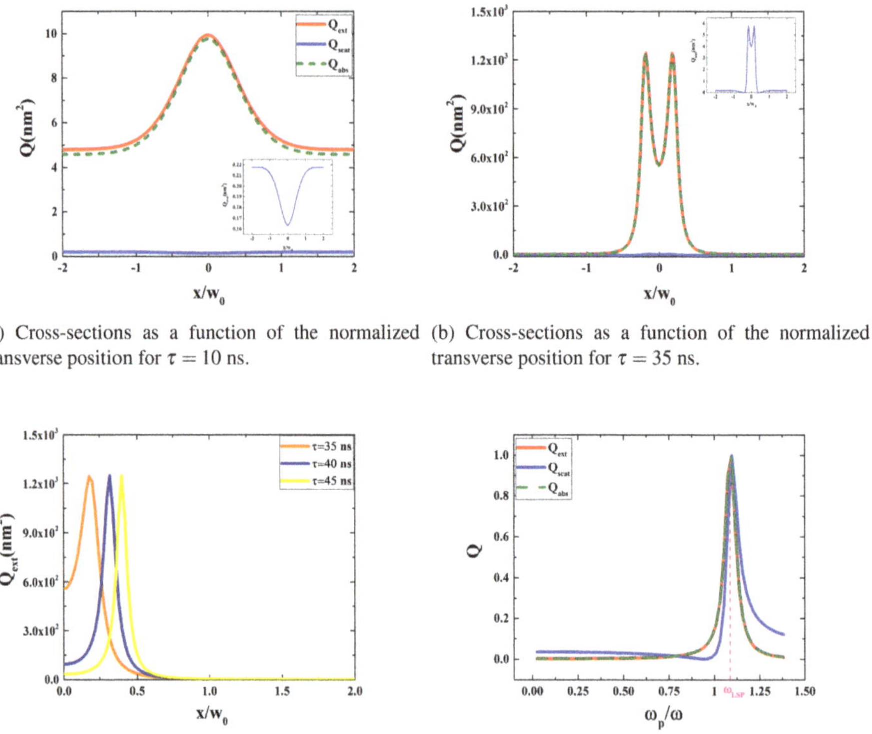

(a) Cross-sections as a function of the normalized transverse position for $\tau = 10$ ns.

(b) Cross-sections as a function of the normalized transverse position for $\tau = 35$ ns.

(c) Extinction cross-section for various values of τ.

(d) Normalized cross-sections as a function of ω_p/ω.

Figure 6.8. Scattering cross-sections of the semiconductor particle. *(a)* Extinction cross-section (solid red line), scattering cross-section (solid blue line), and absorption cross-section (dashed green line) as a function of the normalized transverse position for a semiconductor particle with $a = 15$ nm when $\tau = 10$ ns. Inset: close-up of the scattering cross-section. *(b)* Same as (a) but for $\tau = 35$ ns. Inset: close-up of the scattering cross-section. *(c)* Extinction cross-section as a function of the normalized transverse position for different carrier lifetimes. *(d)* Normalized extinction, scattering, and absorption cross-sections as a function of the plasma frequency normalized by ω. *Parameters:* $a = 15$ nm, $P = 100$ mW; $w_0 = 5$ μm; $\lambda = 1064$ nm; $\mathbf{r}_d = (x/w_0, 0, 0)$. The semiconductor parameters are the same as in figure 6.4.

carriers are generated. Given the carrier density achieved, it is natural for this to be the dominant mechanism.

As the particle approaches the optical axis, i.e., where the local light intensity is higher, both the extinction and absorption cross-sections increase, in agreement with the growth of $\text{Im}\{\Gamma\}$ due to absorption by free carriers. It is also noted that the scattering cross-section (solid blue line), better visualized in the figure inset, decreases as the particle approaches the optical axis, as a result of the reduction in surface charges.

Figure 6.8(b) presents the same scenario, now for $\tau = 35$ ns, highlighting the effect of the LSP resonance. It can be seen that the extinction cross-section remains dominated by absorption, and the LSP resonance manifests as a peak located before the optical axis. At the resonance point, the extinction (and absorption) cross-section reaches a value approximately two orders of magnitude higher than the maximum observed for $\tau = 10$ ns. The scattering cross-section (shown in the inset) also exhibits a maximum at the LSP resonance but remains largely suppressed compared to absorption.

Thus, we conclude that during the interaction of the semiconductor particle with light, absorption is the primary mechanism responsible for the extinction of the incident wave. However, we cannot *a priori* determine the fate of the absorbed energy: it may be converted into heat in the semiconductor's crystal lattice (raising its temperature) or re-emitted as light through radiative recombination processes. The recombination process ultimately dictates the final destination of this energy. Although particle heating can generate photophoretic forces, this topic is beyond the scope of this work. For our purposes, it is sufficient to know that absorption leads to an increase in the charge carrier density. Readers interested in exploring this subject further are referred to [2, 20–22].

The LSP resonance occurs at positions where the induced charge density is sufficient to satisfy $\omega_p = \omega_{LSP}$. Since ω_p depends on the carrier density, the position of the particle relative to the beam where resonance occurs can be controlled by modifying this density, either by varying the carrier generation rate (e.g., adjusting the beam power) or by changing the carrier lifetime. Figure 6.8(c) illustrates this effect by comparing the extinction cross-section of a particle with a radius of 15 nm for three distinct values of τ. It is observed that the longer the carrier lifetime, the farther from the optical axis the particle needs to be to reach the resonance condition. This behavior reflects the fact that increasing τ increases the carrier density for the same light intensity, causing the condition $\omega_p = \omega_{LSP}$ to be met in lower-intensity regions—i.e., farther from the beam center.

The same peak shift is observed in the scattering and absorption cross-sections, as shown in figure 6.8(d). Both exhibit resonance at $\omega_{LSP} = 1.088\ \omega$, confirming that the resonance condition is the same for the different extinction mechanisms.

Unlike metallic plasmonic particles, where the LSP resonance frequency strongly depends on the particle size, in the case of semiconductor particles, this dependence is negligible within the size range where the dipolar approximation is valid. In fact, the description of the cross-sections using equation (2.42), which includes the radiation correction term, is equivalent to the approach used in this chapter for $\lambda = 1064$ nm. This is primarily due to the fact that at this wavelength, $ka \ll 1$.

The generation of carriers by optical absorption imposes a fundamental characteristic on the polarizability of the semiconductor particle: its dependence on the intensity of the Gaussian beam. Since the beam intensity varies in space, the polarizability also exhibits a spatial variation. In other words, the polarization process of the semiconductor particle is nonlinear.

This variation occurs on a scale larger than the particle size. In fact, within the particle volume, both the polarizability and the electric field can be considered uniform in the quasi-static approximation. The relevant spatial dependence occurs on the scale of the beam, $L = w_0$, as discussed in section 2.2.3.

This spatial variation in polarizability modifies the force that the dipole experiences due to electrostatic interaction. As a result, expressions (2.85)–(2.87) are no longer sufficient to fully describe this force, as new terms arise due to the nonlinearity of the system.

To address this effect, we follow the approach[6] of Jiang et *al* [23], who derived the gradient force by considering the interaction between the dipole—with a nonlinear polarizability—and the electric field of a focused Gaussian beam. In Jiang's study, gold nanoparticles (AuNPs) were analyzed under ultrashort laser pulses in the near-infrared, resulting in a nonlinear optical response. In this regime, the electrical permittivity of the particles follows $\varepsilon_r = \varepsilon_0(\chi_1 + \chi_3 I)$, where χ_1 represents the linear susceptibility and χ_3 is the third-order term responsible for the nonlinearity. Analogously to our case, the dielectric constant of the particle depends on the local electric field.

The interaction potential at the centroid position of the particle is given by

$$U_{\mathrm{p}}(\mathbf{r}_{\mathrm{d}}) = -\mathbf{p}(\mathbf{r}_{\mathrm{d}}) \cdot \mathbf{E}(\mathbf{r}_{\mathbf{d}}) = -\varepsilon_0\varepsilon_{\mathrm{m}}\,\mathrm{Re}\{\Gamma(\mathbf{r}_{\mathrm{d}})\}|\mathbf{E}_0(\mathbf{r}_{\mathrm{d}})|^2 = -\frac{1}{2}\frac{n_{\mathrm{m}}}{c}\,\mathrm{Re}\{\Gamma(\mathbf{r}_{\mathrm{d}})\}I(\mathbf{r}_{\mathrm{d}}). \quad (6.14)$$

This potential results from the electrostatic interaction between the induced charges on the particle surface and the electric field of light. Since the generated free carriers act as a shielding mechanism, the induced charge density depends on the local beam intensity, as illustrated in figure 6.7(c). This implies that the variation of the interaction potential occurs not only due to the E^2 term (quadratic dependence on the fields, linear in intensity)—which, as discussed in chapter 2, arises because one of the electric fields induces polarization while the other generates the force (qE)—but also due to carrier generation, which introduces nonlinearity in Γ.

Figure 6.9(a) shows the optical potential generated in a spherical semiconductor particle ($a = 15$ nm) as a function of the normalized transverse displacement for different carrier lifetimes. To understand this behavior, we can compare it with figures 6.7(c) and 6.7(a), which present, respectively, the induced charge density at the particle north pole ($\theta = 0$) and the real part of the polarizability under the same conditions (we use the same color code to represent each carrier lifetime).

For short carrier lifetimes, such as $\tau \leqslant 10$ ns, the free-carrier density is low, and the resulting optical potential (solid black line) remains close to what is expected for a medium without carrier generation (dashed red line), i.e. if $\varepsilon_{\mathrm{r}} = \varepsilon_{\mathrm{opt}}$. This occurs because both the induced charge density and $\mathrm{Re}\{\Gamma\}$ are only slightly affected by the generated free carriers. Consequently, the shielding effect is minimal, and the interaction with the electric field displaces the electronic cloud relative to the nucleus, polarizing the particle. In this case, the optical potential resembles that of a dielectric particle, exhibiting a minimum on the optical axis, which allows for trapping the particle in this position.

As the carrier lifetime increases, the free charge density also grows, enhancing the shielding effect. For $\tau = 20$ ns (solid green line), the optical potential no longer exhibits the characteristics of a dielectric particle. We observe the emergence of a

[6] Although we follow the same general idea, our construction of the interaction potential differs from that of Jiang. While Jiang derives it from the polarization vector of the particle, we construct it based on the induced dipole moment. This approach allows us to include the effect of the medium on the induced charges.

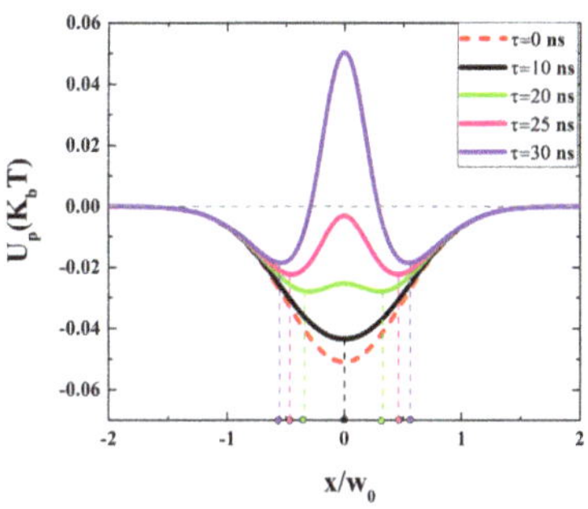

(a) Optical potential in the dielectric regime when $a =$ 15 nm and $w_0 = 5\ \mu m$.

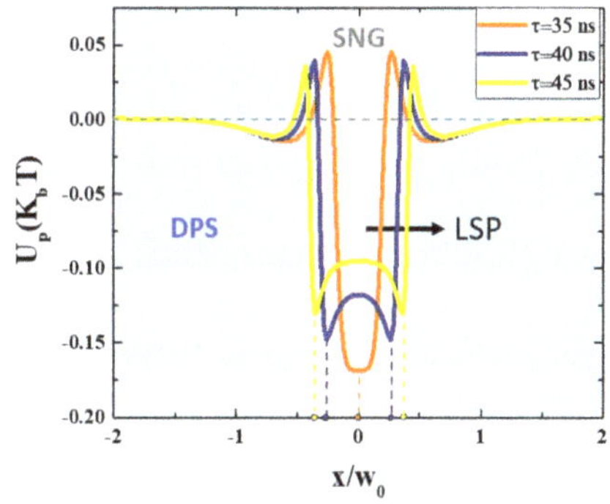

(b) Optical potential in the plasmonic regime when $a =$ 15 nm and $w_0 = 5\ \mu m$.

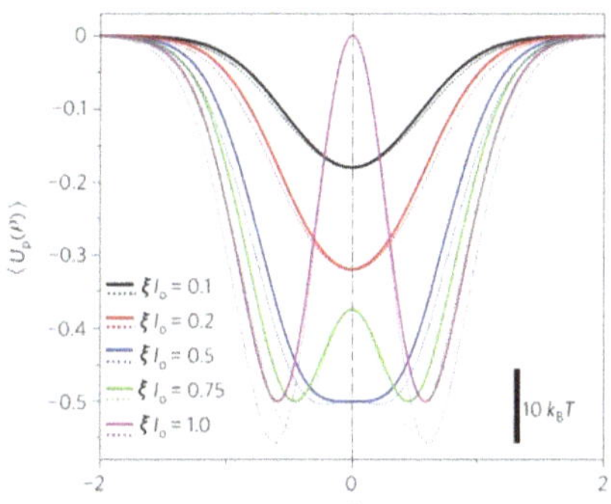

(c) Optical potential in AuNPs when the polarizability is non-uniform.

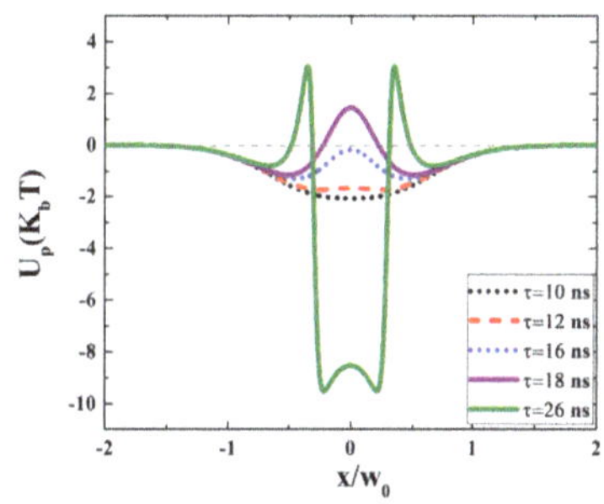

(d) Optical potential in the dielectric regime when $a =$ 50 nm and $w_0 = 4\ \mu m$.

Figure 6.9. Optical potential of the interaction between the particle and the Gaussian beam. *(a)* Optical potential for the semiconductor particle as a function of the normalized transverse position when the semiconductor exhibits the dielectric regime for different carrier lifetimes. $a = 15$ nm, $w_0 = 5\ \mu$m, $P = 100$ mW, $\lambda = 1064$ nm. *(b)* Same as (a) but for carrier lifetimes in which the semiconductor particle exhibits plasmonic behavior. *(c)* Optical potential as a function of radial displacement obtained by Jiang *et al* [23] for AuNPs in a pulsed Gaussian beam focused by an objective with $NA = 0.85$. ξI indicates the strength of the nonlinearity of the AuNPs' polarizability induced by the pulsed beam. $\lambda = 850$ nm, $a = 60$ nm, $P = 150$ mW. Reproduced from [23] with permission from Springer Nature. (d) Same as (a) but with $a = 50$ nm and $w_0 = 4\ \mu$m. *Parameters:* The semiconductor particle parameters are the same as in figure 6.4, $\mathbf{r}_d = (x/w_0, 0, 0)$.

local maximum on the optical axis and the appearance of two lateral minima (green circles on the x-axis of the figure), forming two trapping regions symmetric with respect to the optical axis. This local maximum can be understood by analyzing the behavior of the charge density and the real polarizability: both decrease as the particle approaches the optical axis.

This effect becomes even more pronounced for $\tau = 25$ ns (solid pink line), where the induced charge and the polarizability practically vanish on the optical axis (due to the optical cloaking process). As a consequence, the optical potential exhibits a sharp peak in this region, accompanied by two minima slightly shifted compared to the previous case (pink circles on the x-axis). This means that, as the particle approaches the optical axis, there is an increase in potential energy, making this region unstable. Thus, even though the total potential remains negative, the optical axis no longer represents a stable equilibrium point. This occurs because, as the local intensity increases, the induced charge density decreases, weakening the interaction (potential tends to zero), which disappears when $\varepsilon_r = \varepsilon_m$. The stable equilibrium point shifts to the position where the potential is minimum, which, due to the shielding effect of optically induced free carriers, is no longer where the beam intensity is maximum.

When $\tau = 30$ ns (solid violet line), the induced carrier density becomes sufficient to ensure that $\varepsilon_m \leqslant \varepsilon_r \leqslant -2\varepsilon_m$. This inverts the induced dipole moment, resulting in a negative polarizability, which indicates the induction of negative charges at the particle's north pole. Within this range of ε_r values, both dielectric particles (in the DPS regime) and plasmonic particles (in the DPS and double-negative (DNG) regimes) exhibit a barrier-type potential (positive), as discussed in chapters 2 and 4. This also occurs here for semiconductor particles, where the potential becomes positive in both the dielectric and plasmonic phases, as evidenced by the positive peak near the optical axis. Nevertheless, we observe the existence of two potential minimum positions (when $\varepsilon_r > \varepsilon_m$), corresponding to stable trapping points. These positions are slightly shifted compared to the previous ones (represented by the violet circle on the x-axis).

Although for $\tau = 30$ ns we have $\varepsilon_r < 0$ near the focal axis, the density of induced carriers is still insufficient to excite the LSP resonance in semiconductor particles. However, for $\tau = 35$ ns, this resonance can occur, as shown in the inset of figure 6.7 (c). Figure 6.9(b) displays the optical potential for semiconductor particles at lifetimes where the LSP can be excited. For $\tau = 35$ ns (solid orange line), a broad potential minimum is observed along the optical axis. For longer lifetimes, such as $\tau = 40$ ns (solid dark blue line) and $\tau = 45$ ns (solid yellow line), the potential minimum shifts (yellow and dark blue circles on the x-axis), and the potential exhibits bistability, meaning that the particle can be trapped in two distinct regions. These regions are symmetric with respect to the optical axis, along which the potential presents a local maximum.

This behavior of the optical potential in the plasmonic phase of the semiconductor particle occurs because the induced charge density is maximal at resonance, i.e., when $\varepsilon_r = -2\varepsilon_m$. As the carrier lifetime increases while keeping the beam intensity distribution constant, the relative position to the focus where resonance occurs shifts to regions of lower intensity. If the resonance occurs at $x = a$ and we move the particle to $x < a$, the charge density decreases because we are moving away from resonance, even as the free-carrier density increases, since in this case $\omega_p > \omega_{LSP}$. This weakens the potential (making it more positive), leading to the local maximum on the optical axis. There is a relationship between the stability points in the plasmonic regime and the points where the cross-section is maximized (figure 6.8(c)), indicating the resonance position and confirming the connection between stability regions and the LSP resonance.

The bistability of the optical potential was experimentally observed by Jiang *et al* [23], who managed to trap two 60nm AuNPs in water simultaneously at distinct positions. Figure 6.9(c) shows the optical potential for AuNPs obtained by Jiang using a different approach. As in our model, the potential exhibits bistability, and the equilibrium position shifts with increasing nonlinearity. The potential described by Jiang *et al* is given by $\langle U_p(\rho)\rangle \sim \varepsilon_0(I(\rho) - \xi I(\rho)^2)$, where $\xi = -(3/4)(\chi_3/\chi_1)$ quantifies the intensity of the particle polarization nonlinearity, I is the Gaussian beam intensity, and $\rho = [(x/w_x)^2 + (y/w_y)^2]$, with w_x and w_y representing the beam waists in the x and y directions, respectively. The solid lines show the potential obtained considering only the x-component of the incident electric field (polarization direction), while the dashed lines also include the z-component of the electric field, which arises due to beam focusing by a high numerical aperture objective ($NA = 0.85$).

Although our analysis uses a zeroth-order Gaussian approximation, leading to a more simplified description, our results are qualitatively similar to those of Jiang *et al*, both in the plasmonic and dielectric regimes for semiconductor particles. This strongly suggests that bistability is a characteristic of the particle nonlinear polarizability, allowing for the trapping of multiple particles by exploring points where the potential is minimized. Furthermore, semiconductors emerge as an ideal material for exploring this type of trap, as the effect can be observed in standard OT setups.

In the studied case, where we consider a semiconductor with the optical properties of germanium (with the carrier lifetime as a free parameter), a radius of $a = 15$ nm, and a beam with power $P = 100$ mW and waist $w_0 = 5\ \mu$m, the resulting optical potential is lower than the thermal energy of the medium at room temperature. This indicates that, under these conditions, the potential is not sufficient to trap a semiconductor particle, not even in the plasmonic regime, where the potential is about five times greater. However, by modifying the experimental conditions, for example, by reducing the beam waist and/or increasing the particle size, it is possible to create a viable optical potential. Figure 6.9(d) illustrates this possibility, showing the potential for a particle with $a = 50$ nm and $w_0 = 4\ \mu$m while keeping the other parameters constant. Jiang *et al*'s results also indicate that, in strongly focused beams, the resulting optical potential can be significantly higher than the thermal energy, enabling the use of bistability in the generation of multiple traps.

From the optical potential, we can obtain the gradient force as

$$\mathbf{F}_{\mathrm{Grad}}(\mathbf{r}_\mathrm{d}) = -\nabla U_\mathrm{p}(\mathbf{r}_\mathrm{d}) = \frac{1}{2}\frac{n_\mathrm{m}}{c}\mathrm{Re}\{\Gamma(\mathbf{r}_\mathrm{d})\}\nabla I(\mathbf{r}_\mathrm{d}) + \frac{1}{2}\frac{n_\mathrm{m}}{c}I(\mathbf{r}_\mathrm{d})\nabla[\mathrm{Re}\{\Gamma(\mathbf{r}_\mathrm{d})\}]. \quad (6.15)$$

The first term represents the linear gradient force, structurally identical to equation (2.83), while the second term arises from the nonlinear nature of the polarizability. Figure 6.10 illustrates the behavior of the gradient force in the regimes where the semiconductor particle behaves either as a dielectric or a plasmonic particle.

Figure 6.10(a) presents the x-component of the gradient force as a function of the normalized transverse displacement. To analyze the effect of charge carrier

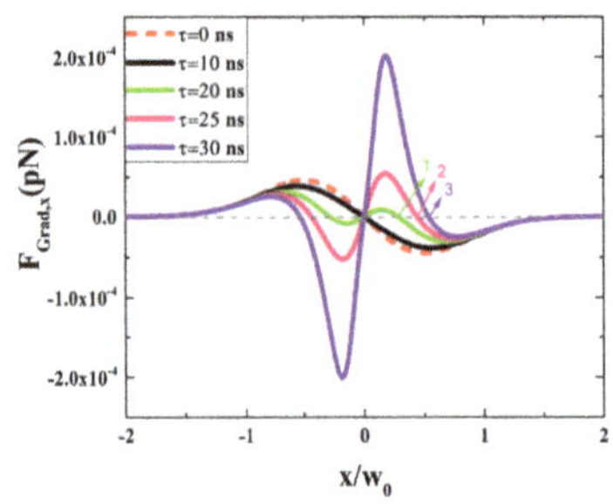

(a) Transverse component of the gradient force in the dielectric regime.

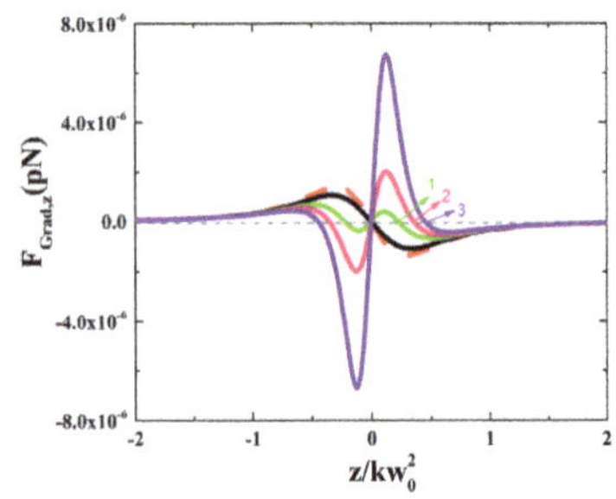

(b) Longitudinal component of the gradient force in the dielectric regime.

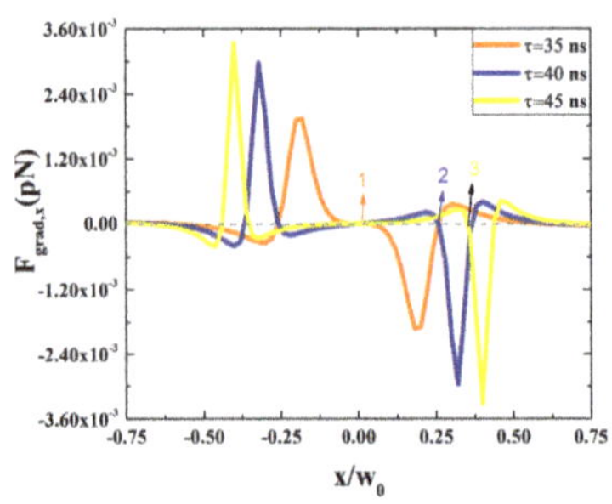

(c) Transverse component of the gradient force in the plasmonic regime.

Figure 6.10. Gradient force on a semiconductor particle. (a) x-component of the gradient force as a function of the normalized transverse position for various carrier lifetimes in which the semiconductor particle behaves as a dielectric. (b) z-component of the gradient force as a function of the normalized longitudinal displacement for various carrier lifetimes in which the semiconductor particle behaves as a dielectric. (c) Same as (a) but for carrier lifetimes capable of inducing plasmonic behavior in the semiconductor particle. The arrows indicate the positions where the particle can be trapped based on the points where the optical potential is minimized. *Parameters:* $a = 15$ nm, $P = 100$ mW, $w_0 = 5\,\mu$m, $\lambda = 1064$ nm, $\mathbf{r}_d = (x/w_0, 0, 0)$ for (a) and (c), $\mathbf{r}_d = (0, 0, z/kw_0^2)$ for (b). The semiconductor particle parameters are the same as in figure 6.4.

generation on the gradient force, we adopt the strategy of varying the lifetime of the photoinduced carriers. In this way, any observed changes in the gradient force result directly from variations in the density of induced carriers rather than modifications in the optical trap itself (such as variations in beam power).

The red dashed line represents the case where no carriers are induced, i.e. $N = N_i$, leading to a constant polarizability. This condition is achieved by assuming $\tau = 0$. Under this assumption, the behavior of the gradient force resembles that of a dielectric particle, allowing for optical trapping along the optical axis, with the force being maximal at $x = \pm w_0/2$.

When the induced carrier density is low and the free-carrier screening effect is negligible, the gradient force remains similar to the dielectric case. This is illustrated by the black solid line, corresponding to $\tau = 10$ ns. However, as the carrier lifetime increases and the density of induced carriers becomes sufficient to significantly reduce the induced charge on the semiconductor particle surface, the gradient force begins to exhibit anomalous behavior. This effect is visible for $\tau = 20$ ns (green solid line), where the trapping position shifts to point 1, and the gradient force becomes positive (repulsive) as the particle approaches the optical axis. This effect becomes even more pronounced for $\tau = 25$ ns (pink solid line), where the trapping position shifts to point 2 (to the right of point 1), and the repulsive nature of the gradient force intensifies.

The transition of the gradient force from attractive (negative) to repulsive (positive) can be intuitively expected, as optical absorption reduces ε_r, and when $\varepsilon_r < \varepsilon_m$, the nature of the gradient force is expected to change. However, this is not the mechanism responsible for the force inversion for $\tau \leqslant 25$ ns. In fact, by analyzing the particle's polarizability (Figure 6.7(a)) and the induced charge density (figure 6.7(c)), we observe that both remain positive, indicating that $\varepsilon > \varepsilon_m$. This demonstrates that the inversion of the gradient force is not directly related to the relationship between the relative permittivity of the particle and the surrounding medium.

This inversion occurs due to the nonlinearity of the polarizability, which induces a local maximum in the optical potential along the optical axis. This behavior can be summarized as follows: '*The gradient force drives the particle toward the region where the optical potential is minimized*'. When the potential minimum occurs on the optical axis, as is the case for dielectric particles when $\varepsilon_p > \varepsilon_m$, the gradient force acts to guide the particle to this region. On the other hand, when the optical potential exhibits a maximum along the optical axis, as in the case where $\varepsilon_p < \varepsilon_m$, the gradient force pushes the particle away from the axis, where the potential is minimized. The behavior of the gradient force, illustrated in figure 6.10(a), directly reflects the behavior of the optical potential shown in figure 6.9(a), confirming that the force always guides the particle toward the potential minimum. This characteristic also explains the observed bistability, as the resulting optical potential exhibits two distinct minima, toward which the gradient force directs the particle.

Naturally, when $\varepsilon_m < \varepsilon_r < -2\varepsilon_m$ and the induced dipole moment becomes antiparallel to the field, the repulsive nature of the gradient force intensifies, as observed for $\tau = 30$ ns (purple solid line). Under this condition, the transition from attractive to repulsive force occurs more abruptly since the resulting optical potential becomes positive. Even in these conditions, there still exist two positions where the particle can be trapped (point 3), where the potential is minimized.

The same behavior is observed in the longitudinal component of the gradient force, as illustrated in figure 6.10(b), and in the plasmonic regime, as shown in figure 6.10(c). In both cases, the gradient force directs the particle toward the potential minimum, which is displaced from the optical axis as the carrier density increases, demonstrating the robustness of this phenomenon.

Within the zero-order Gaussian approximation, only the gradient force acts in the transverse direction. Therefore, for the construction of 2D traps in the transverse plane, the stability condition requires that the generated optical potential be confining and greater than the thermal energy of the medium. According to figure 6.9, we see that this is possible by adjusting the trap conditions or using larger particles.

In the longitudinal direction, on the other hand, in addition to the gradient force, radiation pressure forces also act, originating from scattering and absorption by the particle. As we did in chapter 2, we can treat these effects in terms of extinction. The extinction force accounts for the momentum removed from the electromagnetic wave due to the interaction with the semiconductor particle, both through absorption and scattering of the beam.

In figure 6.8, we show the cross-sections for the semiconductor particle and discuss the effect of nonlinear polarizability on them. In particular, we analyze the shielding caused by optically generated free carriers and its impact on the cross-sections, making them dependent on the local field intensity. Depending on the density of generated carriers, the extinction cross-section may reach its maximum on the optical axis (figure 6.8(a)) or have this maximum shifted to the position where the LSP resonance occurs (figure 6.8(c)).

We should expect these characteristics of the cross-sections to be incorporated into the radiation pressure forces. Indeed, Ghormish *et al* [24] showed that the radiation pressure forces for a particle whose polarizability is nonlinear depend on the intensity of the particle nonlinearity, as they reported for the scattering force on AuNPs when nonlinear effects are induced in them.

To obtain the extinction force, we will use equation (2.92), considering that it is proportional to the time-averaged Poynting vector and to the extinction cross-section, which now becomes position-dependent due to the nonlinearity of the polarizability, as follows,

$$\mathbf{F}_{\text{ext}}(\mathbf{r}_{\text{d}}) = \frac{n_{\text{m}}}{c} Q_{\text{ext}}(\mathbf{r}_{\text{d}}) \langle \mathbf{S}_i(\mathbf{r}_{\text{d}}, t) \rangle = \frac{n_{\text{m}}}{c} Q_{\text{ext}}(\mathbf{r}_{\text{d}}) I(\mathbf{r}_{\text{d}}) \hat{z}. \tag{6.16}$$

The scattering and absorption forces can be directly obtained from equation (6.16). Figure 6.11 illustrates the behavior of the longitudinal forces—the z-component of the gradient force and the extinction force—as a function of the normalized longitudinal displacement. As in previous chapters, we define the resulting optical force in the longitudinal direction as the Rayleigh force ($\mathbf{F}_{\text{Rayleigh}} = \mathbf{F}_{\text{Grad},z} + \mathbf{F}_{\text{ext}}$).

In figure 6.11(a), we show the longitudinal forces acting on the semiconductor particle (a = 15 nm, P = 100 mW, and $w_0 = 5\ \mu$m) when τ = 10 ns, where the behavior is similar to that of a dielectric. It is observed that the extinction force overwhelmingly dominates the gradient force (highlighted in the figure inset), resulting in the acceleration of the particle along the beam propagation direction and preventing the formation of a stable 3D trap. This dominance of the extinction force occurs due to the high optical absorption capacity of germanium at $\lambda = 1064$ nm, reflected in its high imaginary dielectric constant.

To achieve a stable condition, the gradient force must exceed the extinction force. In the dielectric regime, this condition is difficult to obtain, as the maximum extinction force coincides with the region of highest beam intensity. However, in the

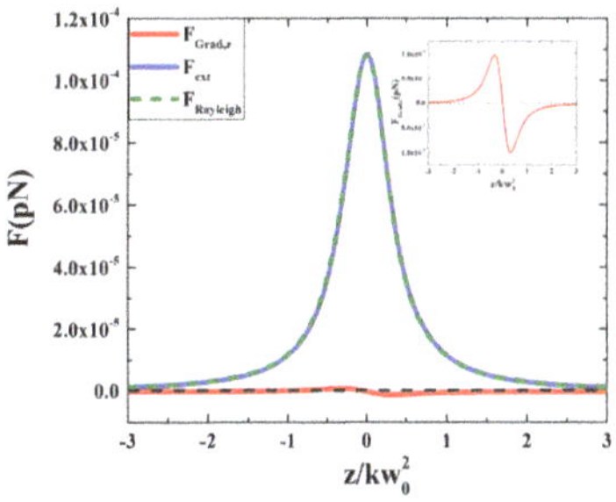

(a) Longitudinal force induced on the semiconductor particle as a function of the normalized longitudinal displacement in the dielectric regime.

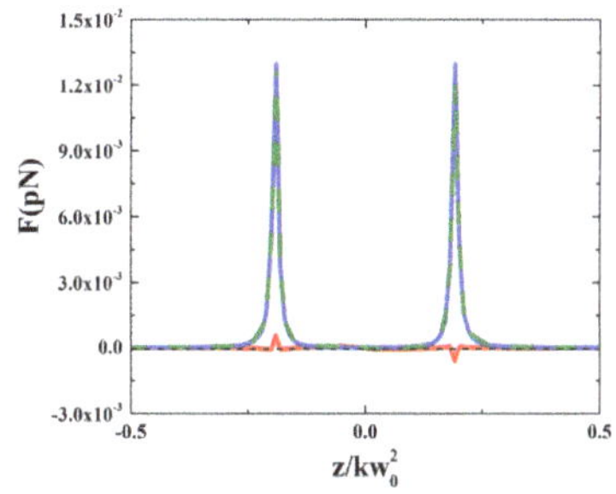

(b) Longitudinal force induced on the semiconductor particle as a function of the normalized longitudinal displacement in the plasmonic regime.

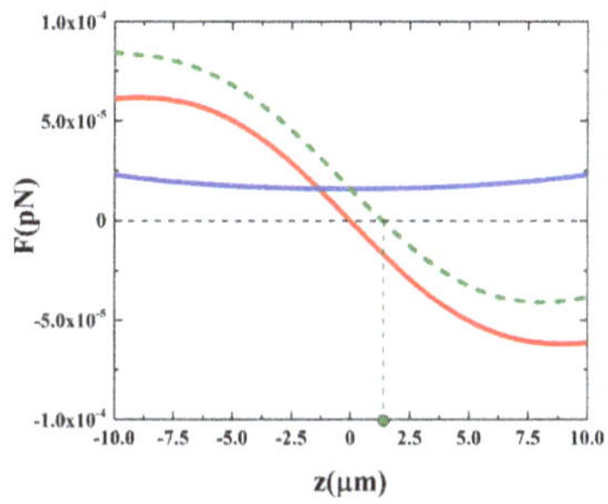

(c) Longitudinal forces induced on the semiconductor particle in the plasmonic regime near the beam waist.

Figure 6.11. Longitudinal forces induced on semiconductor particles as a function of the normalized longitudinal displacement. (a) Longitudinal forces during the dielectric regime. $w_0 = 5\ \mu\text{m}$, $P = 100$ mW, $\tau = 10$ ns. Inset: Highlight of the gradient force. (b) Longitudinal forces during the plasmonic regime. $w_0 = 2\ \mu\text{m}$, $P = 100$ mW, $\tau = 35$ ns. (c) Same as (b), with emphasis on the variation near the beam waist plane ($z = 0$). *Parameters:* $a = 15$ nm, forces were calculated along the optical axis ($x = y = 0$), optical parameters of the particle are the same as in figure 6.4, $\mathbf{r}_\text{d} = (0, 0, z/kw_0^2)$.

plasmonic regime, this stability becomes feasible, as the peak of the extinction cross-section can be shifted to the position of the LSP resonance.

Figure 6.11(b) presents the longitudinal forces when the plasmonic regime is reached with $\tau = 35$ ns. To enhance the gradient force, we reduce the beam waist to $w_0 = 2\ \mu\text{m}$, while maintaining the power at $P = 100$ mW. Similar to the cross-section behavior, the peak position of the extinction force shifts from the highest beam intensity region (optical axis in figure 6.8(c) and beam waist here) to the position of the LSP resonance (figure 6.8(d)). Between these two extinction force peaks, there is a region where the resulting force exhibits longitudinal stability, highlighted in figure 6.11(c), which provides a zoomed-in view of the region near the beam waist ($z = 0$). In this configuration, the z-component of the gradient force surpasses the extinction force for small displacements above and below the beam waist.

Care must be taken when analyzing force (or potential) graphs as a function of displacement for nonlinear particles, since each position is associated with a specific field intensity and, consequently, with distinct values of the optical properties. Therefore, one should not assume that, upon positioning the germanium particle near the beam waist ($z = 0$) and activating the beam, it will be immediately directed to the equilibrium position indicated in figure 6.11(c). The situation illustrated in the graph corresponds to a stationary state, and at the moment the beam is turned on, the particle may undergo a transient regime, in which, for example, it enters LSP resonance and experiences the maximum extinction force before reaching a stable configuration.

References

[1] Ashkin A, Dziedzic J M, Bjorkholm J E and Chu S 1986 Observation of a single-beam gradient force optical trap for dielectric particles *Opt. Lett.* **11** 288

[2] Meyer J R, Kruer M R and Bartoli F J 1980 Optical heating in semiconductors: laser damage in Ge, Si, InSb, and GaAs *J. Appl. Phys.* **51** 5513

[3] Sze S M and Irvin J C 1968 Resistivity, mobility and impurity levels in GaAs, Ge, and Si at 300 K *Solid-State Electron.* **11** 599

[4] Blakemore J S 1982 Semiconducting and other major properties of gallium arsenide *J. Appl. Phys.* **53** R123

[5] Gamaly E G and Rode A V 2014 Transient optical properties of dielectrics and semiconductors excited by an ultrashort laser pulse *J. Opt. Soc. Am.* B **31** C36

[6] Campos W H, Moura T A, Marques O J B J, Fonseca J M, Moura-Melo W A, Rocha M S and Mendes J B S 2019 Germanium microparticles as optically induced oscillators in optical tweezers *Phys. Rev. Res.* **1** 033119

[7] Martinez G *et al* 2017 Determination of the energy band gap of Bi_2Se_3 *Sci. Rep.* **7** 1

[8] Fang M *et al* 2020 Layer-dependent dielectric permittivity of topological insulator Bi_2Se_3 thin films *Appl. Surf. Sci.* **509** 144822

[9] Moura T A, Andrade U M S, Mendes J B S and Rocha M S 2020 Silicon microparticles as handles for optical tweezers experiments *Opt. Lett.* **45** 1055

[10] Davis L W 1979 Theory of electromagnetic beams *Phys. Rev.* A **19** 1177

[11] Barton J P, Alexander D R and Schaub S A 1988 Internal and near-surface electromagnetic fields for a spherical particle irradiated by a focused laser beam *J. Appl. Phys.* **64** 1632

[12] Barton J P and Alexander D R 1989 Fifth-order corrected electromagnetic field components for a fundamental Gaussian beam *J. Appl. Phys.* **66** 2800

[13] Nunley T N, Fernando N S, Samarasingha N, Moya J M, Nelson C M, Medina A A and Zollner S 2016 Optical constants of germanium and thermally grown germanium dioxide from 0.5 to 6.6 eV via a multisample ellipsometry investigation *J. Vac. Sci. Technol.* B **34** 061205

[14] Fox M 2002 Optical Properties of Solids *American Association of Physics Teachers* **70** 1269–70

[15] Gallant M I and Van Driel H M 1982 Infrared reflectivity probing of thermal and spatial properties of laser-generated carriers in germanium *Phys. Rev.* B **26** 2133

[16] Sato S, Ikeda T, Hamada K and Kimura K 2009 Size regulation by bandgap-controlled etching: application to germanium nanoparticles *Solid State Commun.* **149** 862

[17] Tan C-S, Lu M-Y, Peng W-H, Chen L-J and Huang M H 2020 Germanium possessing facet-specific trap states and carrier lifetimes *J. Phys. Chem.* C **124** 13304

[18] Sokolowski-Tinten K and von der Linde D 2000 Generation of dense electron-hole plasmas in silicon *Phys. Rev.* B **61** 2643

[19] Yeh T-T, Shirai H, Tu C-M, Fuji T, Kobayashi T and Luo C-W 2017 Ultrafast carrier dynamics in ge by ultra-broadband mid-infrared probe spectroscopy *Sci. Rep.* **7** 1

[20] Meyer J R, Bartoli F J and Kruer M R 1980 Optical heating in semiconductors *Phys. Rev.* B **21** 1559

[21] Horvath H 2014 Photophoresis–a forgotten force? KONA Powder Part *J* **31** 181

[22] Ambrosio L A, Wang J and Gouesbet G 2022 Towards photophoresis with the generalized Lorenz-Mie theory *J. Quant. Spectrosc. Radiat. Transf.* **288** 108266

[23] Jiang Y, Narushima T and Okamoto H 2010 Nonlinear optical effects in trapping nano-particles with femtosecond pulses *Nat. Phys.* **6** 1005

[24] Mirzaei-Ghormish S, Smalley D and Camacho R 2025 Nonlinear multistable potential traps *Phys. Rev.* A **111** 013514

IOP Publishing

Optical Trapping and Manipulation of New Materials

Tiago de Assis Moura, Joaquim Bonfim Santos Mendes and Márcio Santos Rocha

Chapter 7

Trapping and manipulating particles with intermediate properties: semiconductors and others

We are finally in a position to present in this chapter the general features concerning the optical trapping and manipulation of particles made of materials with intermediate properties between dielectric and conductors. Classical semiconductors such as silicon (Si), germanium (Ge), and others are probably the best example. The general properties of these materials were discussed in detail in chapters 5 and 6. On the other hand, topological insulators, superparamagnetic particles, and particles made of semiconducting polymers are also good examples that can present advantages to be used as tools in some specific applications. We will briefly present the latest findings concerning such materials here, referencing the original works for the reader interested in the full details.

7.1 Why different materials?

In the last few years, many new applications of the optical tweezers (OT) technique were proposed, most of them exploring beams with alternative intensity profiles (i.e. different from the common Gaussian TEM_{00} beams) [1, 2] or, alternatively, exploring the optical properties of new materials [3] that are 'non-usual' for optical trapping experiments such as topological insulators [4], semiconductors [5, 6], and superparamagnetic particles [7–9]. The majority of these applications are based on the fact that the effective characteristics of the optical forces on micro- and nanoparticles can be strongly modulated for specific purposes by exploring the optical properties of the materials employed to construct the particles or by changing the light intensity distribution of the laser profile used in the tweezers, such that one can effectively control the optical trapping and manipulation of the particles.

doi:10.1088/978-0-7503-6074-6ch7

To achieve a good understanding of the emerging scenario above presented, it is fundamental that one understands the dynamics of the effective forces that play a role on particles in OT experiments, a topic extensively discussed in the former chapters. Historically, most studies concerning optical forces were performed for dielectric particles (and some for conducting particles) in past decades. Nevertheless, the literature is still limited on this topic for other types of materials that can be employed to construct beads for OT experiments. In the general cases, both the optical forces (gradient, scattering, along with their peculiarities that strongly depend on the properties of the materials), as well as photophoretic forces [10–15], can compete and play a significant role in the dynamics observed for the particles when in OTs.

7.2 Silicon particles

Parts of this section have been reproduced with permission from [6], copyright 2020 The Optical Society of America.

Si is one of the most important semiconductors used in technological applications, especially in the electronics industry. Therefore, it is natural that one wants to explore this material for novel OT applications.

Recently, the behavior of water-immersed Si beads was investigated in a Gaussian beam (TEM_{00}) OT using a 1064 nm wavelength laser mounted in an inverted microscope with a $100 \times$ N. A. 1.3 oil-immersion objective [6]. These Si particles were produced using the pulsed laser ablation technique [5] (see appendix A), which allows one to obtain microparticles with a regular spherical shape. The shape and composition of the particles were confirmed by scanning electron microscopy and energy dispersive x-ray analysis [6].

Si microparticles (beads) typically tend to perform a 3D oscillatory motion around the focal region of the beam due to the competition of attractive and repulsive forces resultant from their interaction with the laser. Under certain conditions, however, they can also be stably trapped—depending especially on the focus height (distance between the laser focus and the coverslip used to construct the sample chamber).

The study of reference [6] in fact demonstrated for the first time that semiconductor particles can oscillate or can be stably 3D-trapped depending on the focus height, opening the door for the realization of different types of assays with the same bead and the same experimental setup. The oscillatory behavior of semiconductor beads can be useful in OT experiments which are intended to apply dynamical oscillatory forces on systems, like in microrheological studies, in the investigation of the dynamical frequency-dependent response of biopolymers, or in the development of microscopic thermal machines (microscopic engines) [4, 5]. On the other hand, since the trap stiffness achieved when trapping Si beads is about three orders of magnitude smaller than that found for dielectric beads under similar conditions [6], such beads can be used to explore novel applications of OT for working in the femtonewton force range, a feature important in studies concerning weak interactions such as the Casimir forces in colloids [16].

Before the study of reference [6], previous works reported a similar oscillatory behavior for other types of beads with intermediate properties between dielectrics and conductors, including topological insulators [4], other semiconductors [5], and core–shell magnetic beads [9]. Therefore, such behavior can be achieved for a variety of materials and appears to be a general feature of particles made of materials with intermediate properties between dielectric and conductors; verified for a large range of setup parameters including the laser characteristics and the beads sizes.

Figure 7.1(a) shows a schematic drawing representing the oscillatory motion of Si beads close to the optical axis of the tweezers; figure 7.1(b) shows the coordinate frame adopted for force calculations in reference [6] with the origin coinciding with the laser focus; and figure 7.1(c) shows some images of a 2.50 μm radius particle at various positions during its cycle. Beads located in the region below the focal plane ($z < 0$) are attracted to the laser focus. This motion is dominated by the attractive optical gradient forces, which have both transverse (xy-plane) and axial (z-axis) components. As the beads get sufficiently close to the focus they are thus repelled, moving away from this region above the focal plane ($z > 0$), as represented in figure 7.1(a). This change of behavior occurs because a resulting repulsive force starts to dominate the dynamics. Such repulsive force was interpreted in earlier works as composed mainly by at least two components: a photophoretic force [4, 5, 9] and a scattering force [4, 5]. Finally, when sufficiently far from the optical axis the particle tends to sink due to its relatively high density, and the oscillatory cycle restarts.

It is expected that photophoretic forces dominate the resulting repulsive motion when the beads considerably absorb light and heat up at the wavelength used in the tweezers, as in the case of Ge [5] and Fe_3O_4 [9] at TEM_{00} 1064 nm beams, as well as

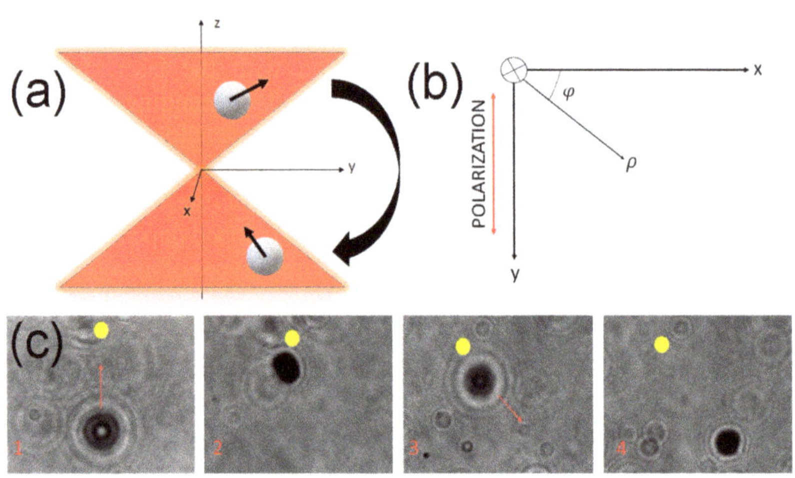

Figure 7.1. (a) Schematic drawing illustrating the 3D oscillatory motion of a Si bead in the tweezers. (b) Reference frame to describe the particle dynamics. (c) Images illustrating the particle motion. The optical axis is marked by a yellow dot here. Adapted with permission from [6], copyright 2020 The Optical Society of America.

of most metallic microparticles [17]. In the case of Si, nevertheless, light absorption at 1064 nm is considerably lower [18] and thus one should ask about the real role of the photophoretic force in the dynamics of these microbeads, as previously commented in chapter 6.

In fact, as previously mentioned (see chapter 6), another relevant effect that can contribute to the total repulsive force is the possibility of occurring plasmonic resonances on the particles as a result of their interaction with the laser light [19, 20]. In the case of the setup described in reference [6], although the laser frequency is fixed ($\omega \sim 1.76 \times 10^{15}$ rad s^{-1}) as in most OT setups, the plasma frequency ω_p of the semiconductor beads can vary because ω_p^2 is proportional to the density of charge carriers N (electrons and holes) in the material [21]. This density changes during the oscillation cycle of the particle due to photon-inducing excitation of electrons that are originally at the valence band of the semiconductor. When photo-excited, such electrons can hop to the conduction band forming an electron–hole pair, therefore changing ω_p. Furthermore, the dielectric constant of the photo-doped semiconductor can be expressed approximately by the formula $\varepsilon_r = \varepsilon(1 - \omega_p^2/\omega^2)$, where ε is the intrinsic dielectric constant of the material [21]. Thus, because $\omega_p^2 \propto N$, the effective dielectric constant (and consequently the refractive index) depends on N. When the particle gets closer to the optical axis in its motion the number of photons per second reaching it considerably increases. Since the electron–hole recombination time is much faster than the particle dynamics, N will depend on the local laser intensity at the particle position [22]. Such an effect induces a decrease on the refractive index of the beads as they approach the focus, and when this quantity becomes smaller than that of the surrounding medium (water) the gradient force inverts its sign, pushing the particle away from the optical axis instead of pulling it to the focus. It is worth saying that this type of phenomenon was also observed for other types of beads [19, 20]. It is noteworthy that such a mechanism was discussed in detail in chapter 6.

One should note that although three different effects (photophoretic force, scattering force and inversion of the gradient force) can contribute to the resulting repulsive force; unfortunately it is very challenging to accurately compare the relative magnitude of these three effects, because such a study would involve accurate measurements of the local dynamical temperature gradient around the particle and the effective changes on its density of charge carriers during the complete oscillation cycle.

In figure 7.2 we show a typical result representing the oscillatory dynamics of a Si bead, denoting its radial position $\rho = \sqrt{x^2 + y^2}$ as a function of the time. Note that there are two very different amplitudes of oscillation, a higher and a smaller one, which depend on the optical properties of the material and on the characteristics of the laser beam, especially its waist. This significant change of the amplitude of oscillation is a feature related to the asymmetry imposed by the linear polarization of the laser beam used to perform the measurements show in figure 7.2 [5]: the beads oscillate with a higher amplitude in a direction parallel to the laser polarization. Nevertheless, Mie scattering theory predicts that there is an

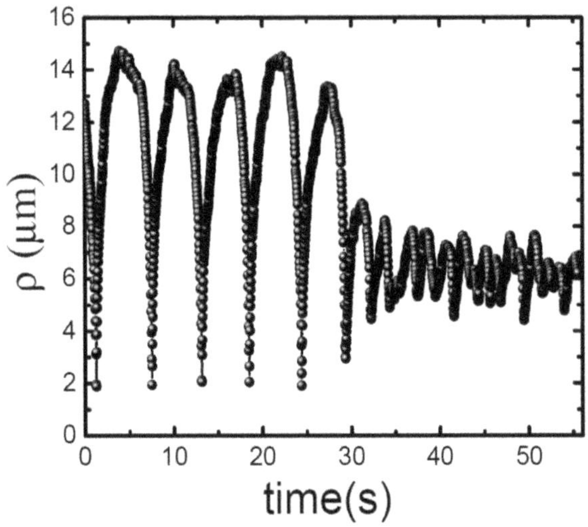

Figure 7.2. Radial position of a 2.50 μm radius Si particle as a function of time. There are two very different amplitudes of oscillation as a consequence of the asymmetry imposed by the laser linear polarization. Adapted with permission from [6], copyright 2020 The Optical Society of America.

azimuthal component of the gradient force that tends to align the particle oscillation direction perpendicularly to the laser polarization [23]. The theory also predicts that the radial component of the gradient force should be higher at the laser polarization direction [23], which explains the decrease of the oscillation amplitude when the particle aligns perpendicularly to the laser polarization [5]. A simple phenomenological model proposed to describe the resultant force (optical + photophoretic components, excluding the Stokes friction on the particle) was proposed in [5]. The ρ-component of this resultant force written in cylindrical coordinates (reference frame in figure 7.1(b)) as the sum of the gradient ($F_{G\rho}$) and photophoretic/scattering ($F_{R\rho}$) components is

$$F_\rho = F_{R\rho} + F_{G\rho} = \mathscr{F}_{R\rho} \exp\left(\frac{-2\rho^2}{w(z)^2}\right) - \frac{2(\rho - \rho_c)\mathscr{F}_{G\rho} \exp(1/2)}{w(z)(1 + \eta_\rho)} \exp\left(\frac{-2\rho^2}{w(z)^2}\right)(1 - \eta_\rho \cos 2\phi), \tag{7.1}$$

where $\mathscr{F}_{G\rho}$ and $\mathscr{F}_{R\rho}$ are the maximum magnitude of the gradient and of the photophoretic/scattering force along the direction ρ, respectively, η_ρ is a parameter that accounts for the relative strength of the asymmetry imposed by the laser linear polarization [23], and $w(z)$ is the beam waist at a distance z from the focal plane. It is related by the beam waist at the focus w_0 by $w(z) = w_0\sqrt{1 + (z\lambda/\pi w_0^2)^2}$ [4]. Finally, ρ_c is the critical position parameter that accounts for a possible inversion of the gradient force at this position, which is expected to be very close to the optical axis [6].

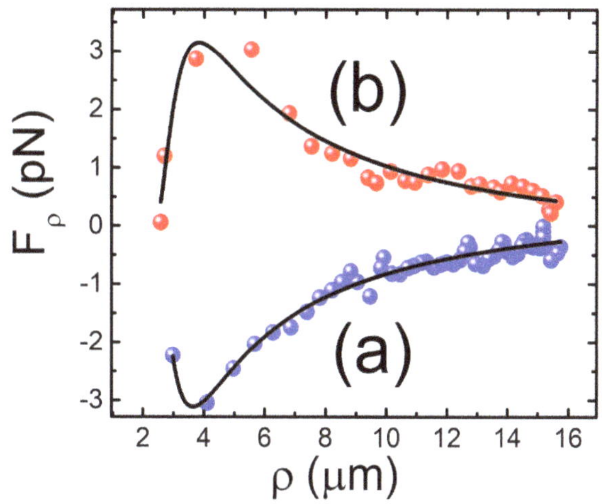

Figure 7.3. Curve (a), *blue*: Radial component of the optical force on a 2.50 μm radius Si particle as a function of its radial distance from the focus ρ for the approach regime. The solid line is a fitting to equation (7.1). Curve (b), *red*: Equivalent result for the removal regime. Adapted with permission from [6], copyright 2020 The Optical Society of America.

Figure 7.3 illustrates the typical behavior of the resultant radial force as a function of ρ for the approach (curve (a), *blue online*) and removal (curve (b), *red online*) motions. Fittings performed using equation (7.1) are also shown as solid lines, which return values for the force amplitudes (coupled with the asymmetry parameter) and for beam waist [6]. These fittings also allow the determination of ρ_c, which is nearly zero for the approach regime but ~3.7 μm for the removal regime. Since ρ_c corresponds to the position where the gradient force invert its sign, the fittings allow one to conclude that such inversion occurs only for the removal regime. On the other hand, the repulsive (photophoretic/scattering) component of equation (7.1) is always positive in both regimes, as expected.

The behavior of the optical force corresponding to the approach regime (figure 7.3 curve (a), *blue*) is very similar to that found for Ge and bismuth telluride beads [4, 5] under similar experimental conditions. The force curve measured for the repulsive force (figure 7.3 curve (b), *red*), on the other hand, is qualitatively different from the corresponding curves reported for these other materials: for Si, the resultant force in the removal regime rapidly increases until $\rho \sim \rho_c$ and then slowly decreases. For Ge and bismuth telluride only a decrease of the force as a function of ρ was found, which is compatible to an always negative gradient force during the oscillatory cycle [4, 5]. This comparison suggests that the inversion of the gradient force really plays a role in the oscillatory dynamics of the Si beads, associated with the low light absorption of this material at 1064 nm, which tends to weaken the photophoretic contribution in the total repulsive force [6].

By recording and analyzing the oscillatory motion of the beads using video-microscopy, various important quantities (amplitude and frequency of the motion, for example) can be measured as a function of parameters of interest (laser power, focus

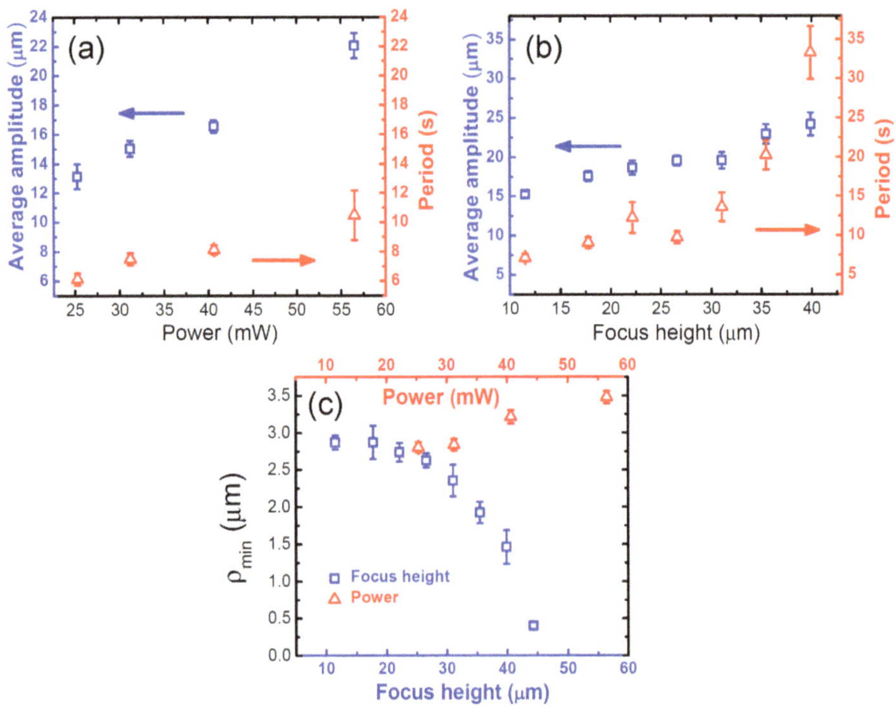

Figure 7.4. Important quantities as functions of setup parameters for the typical dynamics of a Si bead in a TEM_{00} (Gaussian beam) OT with 1064 nm wavelength. Panel (a): average amplitude (parallel to laser polarization) and period of the motion as a function of the laser power at the objective entrance. Panel (b): same parameters of a function of the focus height for a fixed laser power. Panel (c): nearest position that the particle gets to the focus as a function of the laser power and focus height. Adapted with permission from [6], copyright 2020 The Optical Society of America.

height, bead radius, etc), allowing a robust characterization of the dynamics [6]. Figure 7.4 illustrates some of the results reported in reference [6], showing various important quantities as functions of setup parameters for the typical dynamics of a Si bead in a TEM_{00} (Gaussian beam) OT with 1064 nm wavelength.

In reference [6], stable trapping of Si beads was also verified for high focus heights (relatively to the coverslip surface), >43 μm, indicating that a transition between the two regimes (stable trapping and oscillatory motion) is possible depending on the setup parameters. Interestingly, in this case stable trapping was achieved with very low values for the trap stiffness—on the order of fN μm^{-1}, which is about three orders of magnitude smaller than the stiffness typically measured for dielectric beads under similar conditions [6]—a feature that can be explored in experiments in which one intends to apply femtonewton forces on the systems of interest, as commented before. Furthermore, the equilibrium position verified for stable trapping was found to be out of the optical axis, indicating a delicate balance between attractive and repulsive forces [6].

Finally, it is worth commenting that the oscillatory motion of micrometer-sized beads in OT was not known until very recently (2017), but is now being reported for a gamma of particles made of materials with intermediate properties between dielectric and conductors. As perspectives for exploring these systems in the future we cite the development of micromotors and single-molecule machines [4, 5, 24, 25] and the application of controllable oscillatory forces in microrheology [26] and single-molecule force spectroscopy [27, 28] assays.

7.3 Germanium particles

Despite being discovered over a century ago, Ge remains a material of significant interest across condensed matter physics and materials science. Its spin relaxation suppression properties have made Ge and other group IV semiconductors, such as Si, attractive for spintronics research [29–33]. Furthermore, Ge's superior carrier mobility compared to Si [32] has spurred renewed investigation by the solid-state device community to enhance transistor performance [33]. Ge's applications extend to optical fibers, polymerization catalysts, and Si–Ge alloys in microchip manufacturing, where feature sizes are now reaching 7 nm [33, 34]. However, the optical trapping of Ge microparticles had not been explored until recently.

The dynamics of Ge particles is qualitatively similar to that verified for Si particles discussed in the former section, although here the dynamics is faster and the photophoretic forces tend to play a more important role due to the strong absorption of Ge at 1064 nm [5]. A significant difference is that for Ge stable trapping was not verified when using a Gaussian (TEM_{00}) beam, but only when using Bessel beams [35], a feature that evidences the stronger character of the photophoretic forces associated with Gaussian beams in this case [35].

Figure 7.5 synthesizes the results reported in reference [35] concerning the different behaviors verified for Ge beads in highly focused Bessel beam OT; a

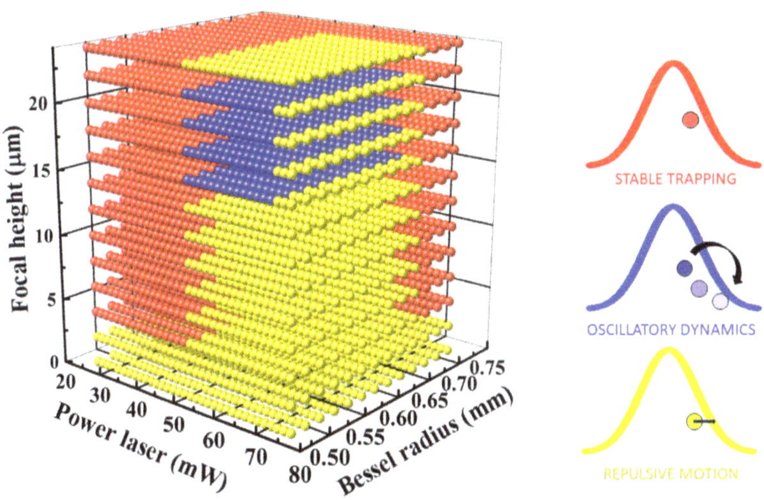

Figure 7.5. 'Phase diagram' showing the different behaviors of a 2.48 μm radius Ge bead in a Bessel beam OT, verified as a function of some parameters of interest: the Bessel radius of the beam, the laser power (at the objective back aperture) and the focal height. *Blue points*: oscillatory dynamics. *Red points*: stable trapping. *Yellow points*: repulsive motion. Reprinted from [35], copyright 2023, with permission from Elsevier.

'phase diagram' that shows the different situations for a set of relevant parameters of interest: the Bessel radius of the beam, the laser power and the focal height (distance from the focal point to the coverslip surface). Figure 7.5 was obtained using a 2.48 μm radius Ge bead, although the general behavior is qualitatively similar for other sizes. Observe that there is a limited region of parameters for which the oscillatory motion occurs in this case (*blue points* in the figure). Furthermore, higher values of the focal height in general favor stable trapping (*red points*) or the oscillatory motion (*blue points*) of the Ge beads, while lower values of this parameter favor only a repulsive motion (*yellow points*) related to dominant repulsive forces. This last behavior is also favored for higher values of the laser power and lower values of the Bessel radius due to a tendency of the Ge bead to absorb more power under these situations [35].

These results explicitly show the high versatility of Ge microbeads to be used as tools in OT experiments. In fact, by simply adjusting setup parameters one can easily change the dynamics of the beads close to the focal region of the laser: they can be stably trapped, oscillate or even be repelled from the focus. The range of applications for the near future is promising.

7.4 Fluorescent polymer nanoparticles (pdots)

Another type of new material recently studied and characterized for OT experiments are fluorescent nanoparticles (pdots) made from the polymer poly[2-methoxy-5-(2'-ethylhexoxy)-*p*-phenylene vinylene] (MEH-PPV) [36]. Although the optical trapping and manipulation of similar nanostructures such as quantum dots was previously reported by different groups [37–39], the work of reference [36] pioneered the use of optical force models that predict the trap stiffness of the nanoparticles. This modeling returns results that, when fitted to the corresponding experimental data, allow one to determine the optical properties of these nanoparticles, such as the refractive index and their optical anisotropy. Moreover, since the MEH-PPV pdots absorb at different wavelengths, these properties—especially the refractive index—change for different wavelengths used to illuminate the sample; and the authors were able to measure and discriminate these changes with accuracy [36]. In other words, this approach allows one to determine the optical properties of materials at the nanoscale, an issue that is not straightforward.

The model of reference [36] predicts that the resulting transverse component (radial component ρ) of the gradient force on the nanoparticles is

$$F_\rho = \frac{-2n}{c\varepsilon_{\mathrm{m}}(1+\eta)}\mathrm{Re}(\alpha_{\mathrm{dyn}})\frac{\rho P}{\pi\omega^4(z)}\exp\left(\frac{-2\rho^2}{\omega^2(z)}\right)[1-\eta\cos(2\theta)], \tag{7.2}$$

where $\theta = \phi$ for κ_y and $\pi/2$ - ϕ for κ_x (x and y are two axis perpendicular to the laser propagation direction z), n is the refractive index of the particle (distinct for x and y due to the anisotropy of the material), c is the light speed in vacuum, P is the laser power at the focus, and $\omega(z)$ is the beam waist at a distance z from the focal plane. An asymmetry factor $1 - \eta\cos(2\theta)$, related to Mie resonance effects on the particle, was taken into account due to the linear polarization of the laser used in this case [5, 23]. All the other parameters of equation (7.2) were defined previously.

With this model, the trap stiffness could be calculated and compared to experimental data.

Figure 7.6 shows the main results obtained in reference [36]. Observe that the model predicts very well how the trap stiffness changes with the laser power and with the pdot radius, allowing one to determine the refractive indices and anisotropy parameter for different wavelengths used to illuminate the sample. Table 7.1 depicts the results obtained by the authors.

In summary, the work of reference [36] proved that is possible to use optical force models, even simplified models, to fit experimental data of the trap stiffness and determine important optical parameters of the trapped particles with sizes in the nanoscale range. In the future, this type of approach has potential to become a standard in the characterization of nanomaterials, especially with the use of more sophisticated optical force models.

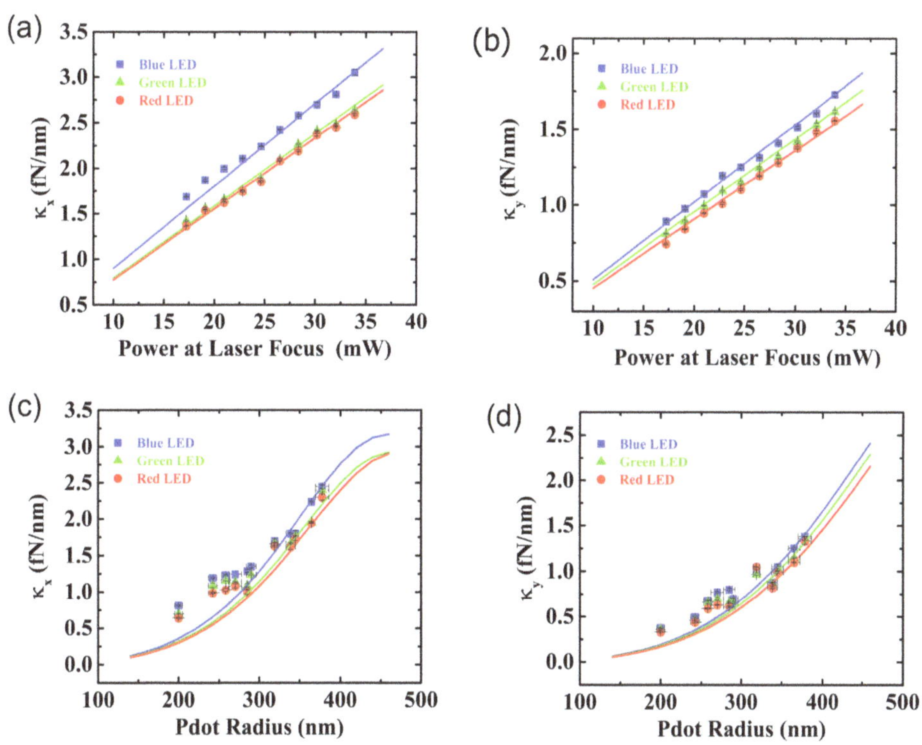

Figure 7.6. (a) and (b) Transverse trap stiffness measured along two perpendicular directions (κ_x, κ_y) for a 365 nm radius pdot (at 10 μm above the bottom of the sample chamber) as a function of the laser power at the focus, for the three different illumination wavelengths used. (c) and (d) Same quantities for different pdots as a function of their radius, for a fixed laser power of 25 mW at the focus. Reprinted figure with permission from [36], copyright 2024 by The American Physical Society.

Table 7.1. Fitting parameter results obtained for the refractive indices and asymmetry parameters at the three wavelengths used to illuminate the sample. The maximum error bars estimated were <2.2% for the refractive indices and <2.5% for the asymmetry parameters. These errors were estimated by performing a statistical analysis considering the experimental uncertainty on the waist of the laser beam; the error associated with the fitting process is much smaller. Reprinted table with permission from [36], copyright 2024 by The American Physical Society.

LED	n_x	n_y	η
Blue	1.428	1.382	0.03
Green	1.423	1.379	0.07
Red	1.421	1.376	0.07

7.5 Magnetic microparticles

Magnetic particles can be constructed for various applications using iron oxide and a dielectric matrix, for example, iron oxide dots embedded in the bulk of a polystyrene bead or a core–shell particle composed of polystyrene coated by an iron oxide shell. In the first case, the iron oxide dots usually present sizes smaller than the size of the magnetic domains of the material, which results in *superparamagnetic* beads—the ideal tools for magnetic tweezers experiments due to their lack of hysteresis. In any case, the optical trapping and manipulation of such beads can be challenging because of the high absorption of light by the iron content, which leads to strong repulsive photophoretic forces.

Recently, Zhong *et al* [9] reported an oscillatory dynamics for core–shell magnetic particles of diameters between 4.0 and 5.0 μm, constructed using an iron oxide shell coating a polystyrene bead. Such oscillatory motion is similar to the dynamics reported here before for Si and Ge beads and was also observed in Gaussian (TEM_{00}) tweezers with wavelength 1064 nm. The authors interpreted the dynamics observed as a result of the competition between gradient forces and photophoretic forces [9], similarly to what was discussed in the previous sections.

Superparamagnetic beads constructed using iron oxide dots embedded in a polystyrene matrix, on the other hand, have not exhibited this type of oscillatory motion, being simply expelled from the focal region of Gaussian beam tweezers due to the dominant photophoretic forces [8]. When using a highly focused Bessel beam, however, Andrade *et al* were able to stably trap and manipulate such beads, opening the door for the development of hybrid instruments such as optomagnetic tweezers, capable of simultaneously applying optical and magnetic forces on beads [8]. This work was already presented briefly in chapter 3, section 3.4. In this case, a geometrical optics approach considering light absorption was able to explain the optical forces on a 2.8 μm radius superparamagnetic bead, presenting good agreement with experimental data (see section 3.4 and/or reference [8]).

7.6 Organic semiconductor particles

The use of the organic semiconductor polyaniline (PANI) for the preparation of spherical-shaped microparticles to serve as handles in OT experiments was also recently proposed [40].

Micrometer-sized PANI particles could be stably trapped, and the optical forces could be also well described by a geometrical optics (GO) approach as in the former case (superparamagnetic particles) [40]. No oscillations were observed here under the experimental conditions used by the authors (TEM_{00} beam of 1064 nm wavelength; PANI beads in water). Figure 7.7 shows the main results obtained in reference [40] for the optical trapping of PANI microparticles.

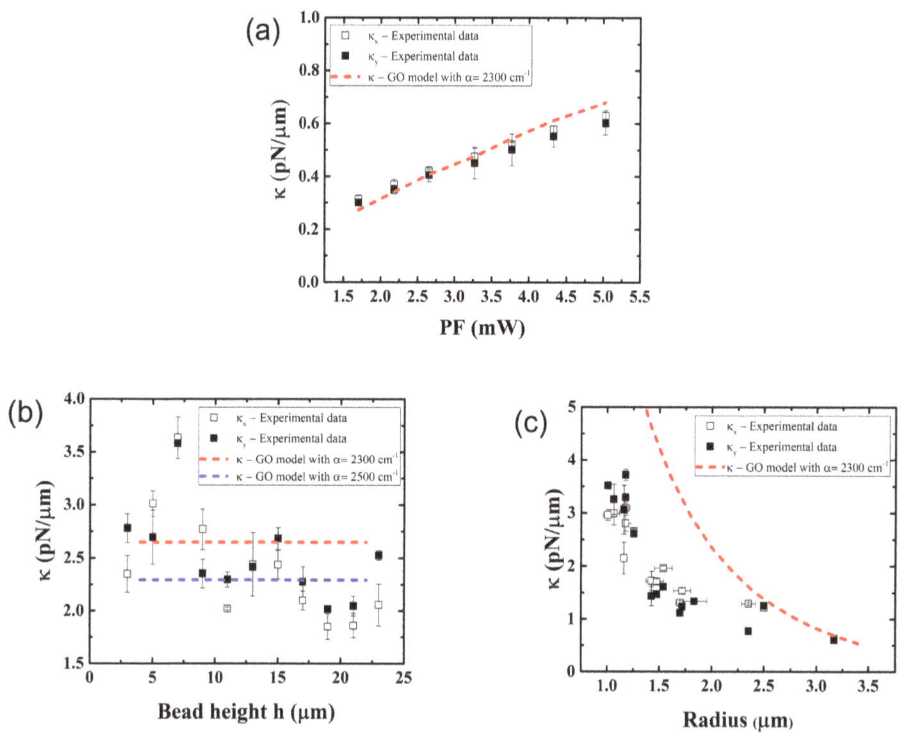

Figure 7.7. Panel (a): trap stiffness (measured in two perpendicular directions x and y) as a function of the laser power at the focus for a 3.17 μm radius PANI particle located 10 μm above the coverslip of the sample chamber. The dashed line is the prediction of the geometrical optics model presented in chapter 3 considering an absorption coefficient of 2300 cm^{-1}. Panel (b): trap stiffness as a function of the bead height relative to the coverslip surface for a 2.50 μm radius PANI bead with a fixed laser power, presenting a good agreement with the GO model. Panel (c): trap stiffness as a function of the bead radius for a fixed height and laser power. In this case the deviations are larger, which was interpreted in reference [40] as a limitation of the GO model to describe the optical forces when there are significant absorption and changes in the absorption coefficient of the beads (fixed in the model calculations). Reprinted with permission from [40], copyright 2023 American Chemical Society.

Panel (a) shows the trap stiffness (measured in two perpendicular directions x and y) as a function of the laser power at the focus for a 3.17 μm radius PANI particle located 10 μm above the coverslip of the sample chamber. The dashed line is the prediction of the geometrical optics model presented in chapter 3 considering an absorption coefficient of 2300 cm^{-1}. The agreement between the model and the experimental data is good, but observe that there are non-negligible deviations. As discussed in reference [40], for the lowest powers the experimental data is slightly higher than the theoretical prediction, and the opposite occurs for the highest powers used. Such behavior suggests that the absorption coefficient of PANI beads is intensity-dependent, as in the case of traditional semiconductors such as Ge and Si [40].

Figure 7.7(b) shows the trap stiffness as a function of the bead height relative to the coverslip surface for a 2.50 μm radius PANI bead with a fixed laser power, presenting a good agreement with the GO model. It is worth commenting, however, that the experimental data presents a slight decrease while the GO model predicts a constant trap stiffness as a function of the bead height—a consequence of the fact that the model used in reference [40] does not consider any optical aberration.

Finally, figure 7.7(c) shows the trap stiffness as a function of the bead radius for a fixed height and laser power. In this case the deviations are larger, which was interpreted in reference [40] as a limitation of the GO model to describe the optical forces when there are significant absorption and changes in the absorption coefficient of the beads (fixed in the model calculations).

References

[1] Curtis J E, Koss B A and Grier D G 2002 Dynamic holographic optical tweezers *Opt. Commun.* **207** 169–75

[2] Dufresne E R, Spalding G C, Dearing M T, Sheets S A and Grier D G 2001 Computer-generated holographic optical tweezer arrays *Rev. Sci. Instrum.* **72** 1810

[3] Skelton Spesyvtseva S E and Dholakia K 2016 Trapping in a material world *ACS Photon* **3** 719

[4] Campos W H, Fonseca J M, Carvalho V E, Mendes J B S, Rocha M S and Moura-Melo W A 2018 Topological insulator particles as optically induced oscillators: Toward dynamical force measurements and optical rheology *ACS Photon* **5** 741

[5] Campos W H, Moura T A, Marques O, Fonseca J M, Moura-Melo W A, Rocha M S and Mendes J B S 2019 Germanium microparticles as optically induced oscillators in optical tweezers *Phys. Rev. Res.* **1** 033119

[6] Moura T A, Andrade U M S, Mendes J B S and Rocha M S 2020 Silicon microparticles as handles for optical tweezers experiments *Opt. Lett.* **45** 1055

[7] Oliveira L, Campos W H and Rocha M S 2018 Optical trapping and manipulation of superparamagnetic beads using annular-shaped beams *Methods Protoc.* **1** 44

[8] Andrade U M S, Garcia A M and Rocha M S 2021 Bessel beam optical tweezers for manipulating superparamagnetic beads *Appl. Opt.* **60** 3422

[9] Zhong M-C, Liu A-Y and Ji F 2019 Opto-thermal oscillation and trapping of light absorbing particles *Opt. Exp.* **27** 29730

[10] Mehta K K, Wu T-H and Chiou E P Y 2008 Magnetic nanowire-enhanced optomagnetic tweezers *Appl. Phys. Lett.* **93** 254102

[11] Iyengar S S, Praveen P, Rekha S, Ananthamurthy S and Bhattacharya S 2014 Trapping characterization of semi metallic magnetic beads in optical tweezers *OSA Photon.* **T3A** 40
[12] Rohatschek H 1985 Direction, magnitude and causes of photophoretic forces *J. Aeros. Sci.* **16** 29
[13] Zulehner W and Rohatschek H 1995 Representation and calculation of photophoretic forces and torques *J. Aeros. Sci.* **26** 201
[14] Horvath H 2014 Photophoresis–a forgotten force? *KONA Powder Part. J.* **31** 181
[15] Ambrosio L A, Wang J and Gouesbet G 2022 Towards photophoresis with the generalized Lorenz–Mie theory *J. Quant. Spec. Radiat. Transf.* **288** 108266
[16] Ether D S *et al* 2015 Probing the casimir effect with optical tweezers *Europhys. Lett.* **112** 44001
[17] Ke P C and Gu M 1999 Characterization of trapping force on metallic mie particles *Appl. Opt.* **38** 160
[18] Lank N O, Johansson P and Kall M 2018 *Opt. Exp.* **26** 29074
[19] Arias-Gonzalez J R and Nieto-Vesperinas M 2003 *J. Opt. Soc. Am.* A **20** 1201
[20] Lehmuskero A, Johansson P, Rubinsztein-Dunlop H, Tong L and Kall M 2015 *ACS Nano* **9** 3453
[21] Fox M 2010 *Optical properties of solids.* (Oxford: Oxford University Press)
[22] Gallant M I and van Driel H M 1982 Infrared reflectivity probing of thermal and spatial properties of laser-generated carriers in germanium *Phys. Rev.* B **26** 2133
[23] Dutra R S, Viana N B, Maia Neto P A and Nussenzveig H M 2007 *J. Opt. A: Pure Appl. Opt.* **9** S221
[24] Zou X, Zheng Q, Wu D and Lei H 2020 Controllable cellular micromotors based on optical tweezers *Adv. Func. Mat.* **30** 2002081
[25] Li J, Kollipara P S, Liu Y, Yao K, Liu Y and Zheng Y 2022 Opto-thermocapillary nanomotors on solid substrates *ACS Nano* **16** 8820
[26] Evans Manlio Tassieri R M L, Warren R L, Bailey N J and Cooper J M 2012 Microrheology with optical tweezers: data analysis *New J. Phys.* **14** 115032
[27] Oliveira L and Rocha M S 2017 Force spectroscopy unravels the role of ionic strength on dna-cisplatin interaction: Modulating the binding parameters *Phys. Rev.* E **96** 032408
[28] Rocha M S 2009 Modeling the entropic structural transition of dna complexes formed with intercalating drugs *Phys. Biol.* **6** 036013
[29] Pezzoli F *et al* 2012 Optical spin injection and spin lifetime in ge heterostructures *Phys. Rev. Lett.* **108** 156603
[30] Santos E, Abrão J E, Costa J L, Santos J G S, Rodrigues-Junior G, Mendes J B S and Azevedo A 2024 Negative orbital hall effect in germanium *Phys. Rev. Appl.* **22** 064071
[31] Bottegoni F *et al* 2017 Spin-hall voltage over a large length scale in bulk germanium *Phys. Rev. Lett.* **118** 167402
[32] Sze S M and Ng K K 1981 *Physics of Semiconductor Devices* 2nd edn (New York: Wiley)
[33] Pillarisetty R 2011 Academic and industry research progress in germanium nanodevices *Nature* **479** 324
[34] Burdette S C and Thornton B F 2018 The germination of germanium *Nat. Chem.* **10** 244
[35] Moura T A, Andrade U M S, Mendes J B S and Rocha M S 2023 Modulating the trapping and manipulation of semiconductor particles using bessel beam optical tweezers *Optics Lasers Eng.* **170** 107778

[36] Moura T A, Lana Júnior M L, da Silva C H V, Américo L R, Mendes J B S, Brandão M C N P, Subtil A G S and Rocha M S 2024 2024 Optical trapping and manipulation of fluorescent polymer-based nanostructures: Measuring optical properties of materials in the nanoscale range *Phys. Rev. Appl.* **22** 064043
[37] Maragò O M, Jones P H, Gucciardi P G, Volpe G and Ferrari A C 2013 Optical trapping and manipulation of nanostructures *Nat. Nanotech.* **8** 807
[38] Sudhakar S, Abdosamadi M K, Jachowski T J, Bugiel M, Jannasch A and Schäffer E 2021 Germanium nanospheres for ultraresolution picotensiometry of kinesin motors *Science* **371** eabd9944
[39] Kolbow J D, Lindquist N C, Ertsgaard C T, Yoo D and Oh S-H 2021 Nano-optical tweezers: Methods and applications for trapping single molecules and nanoparticles *Chem. Phys. Chem.* **22** 1409
[40] Oliveira K M, Moura T A, Lucas J L C, Teixeira A V N C, Rocha M S and Mendes J B S 2023 Use of organic semiconductors as handles for optical tweezers experiments: Trapping and manipulating polyaniline (pani) microparticles *ACS Appl. Polym. Mater.* **5** 3912

IOP Publishing

Optical Trapping and Manipulation of New Materials

Tiago de Assis Moura, Joaquim Bonfim Santos Mendes and Márcio Santos Rocha

Appendix A

Methods of synthesis of semiconductor particles

In this appendix, methods for the preparation of micrometer-sized particles from semiconductor crystals will be discussed. An efficient procedure consists in preparing the particles with the use of laser ablation synthesis in liquid solution (LASLS) from crystals of commercially purchased semiconductor materials. As a sustainable and 'green' method, LASLS enables the creation of diverse nanomaterials [1]. A single experimental setup enables the synthesis of diverse nanostructures, including noble metal spheres, multiphase core–shell oxides, metal—semiconductor heterostructures, and layered organometallic compounds, simply by adjusting a few synthesis parameters. Here, we discuss the procedures for synthesizing particles from semiconductor materials using the LASLS technique, which made it possible to carry out the optical tweezers experiments with these particles discussed in chapter 7.

A.1 Synthesis and characterization of the germanium (Ge) and silicon (Si) microparticles

As mentioned previously, pulsed laser ablation in liquid solution can be employed to synthesize Ge and Si microparticles with the purpose of carrying out experiments with optical tweezers [2, 3]. Specifically, for the synthesis of the Ge particles discussed in chapter 7, this ablation process was conducted with a Nd:YAG laser Quantel, model Brilliant B, employing its second harmonic output at a wavelength of 532 nm. The laser pulses, delivered in the nanosecond regime, have an energy of 70 mJ and a repetition rate of 10 Hz. The experimental setup used to synthesize the microparticles consists of using high purity Ge crystals (99.999%) as a target material immersed in 10 ml of distilled water as a working liquid for ablation. The target was ultrasonically cleaned in distilled water before the process and placed at the bottom of a beaker. The incident laser beam, emanating from the source, was meticulously directed and focused onto the surface of the Ge target, which was submerged beneath a layer of distilled water approximately 10.0 mm in thickness, thereby creating a controlled liquid environment; this focusing was achieved through

doi:10.1088/978-0-7503-6074-6ch8

the utilization of a 50 mm focal length lens, resulting in a focused spot size of approximately 1.0 m in diameter on the target's surface. To ensure a consistent and uniform ablation process, and to prevent localized depletion of the target material, the Ge target was subjected to a controlled translational motion, specifically a movement perpendicular to the path of the incident laser beam, for 5 min, thereby continuously exposing fresh, un-ablated surfaces to the laser irradiation throughout the entirety of the experimental procedure. Finally, all experimental procedures were conducted under ambient conditions, specifically at room temperature, which was measured to be approximately 295 K, and under standard atmospheric pressure, equivalent to 1 atm.

Si particles were synthesized in a similar manner using the second harmonic ($\omega = 532$ nm) with a pulse energy of 50 mJ and 10 Hz in the nanosecond regime. The laser was focused on the surface of the Si target (purity 99.99%) under a liquid layer (distilled water) of approximately 10 mm thickness with a spot of about 1 mm in diameter using a 50 mm focal lens. The target was moved perpendicular to the laser beam for 10 min to irradiate fresh surfaces throughout the process. The experiments were conducted at room temperature ($T \simeq 295$ K) and a pressure of 1 atm.

Scanning electron microscopy (SEM) and Raman spectroscopy analyses were employed to confirm the high quality and structural integrity of the semiconductor microparticles. Figure A.1 shows scanning electron microscopy results of Ge beads obtained by the laser ablation technique. Scanning electron microscopy images demonstrate a size distribution for the laser-ablated particles, ranging from hundreds of nanometers up to a few microns in diameter. Note that the SEM

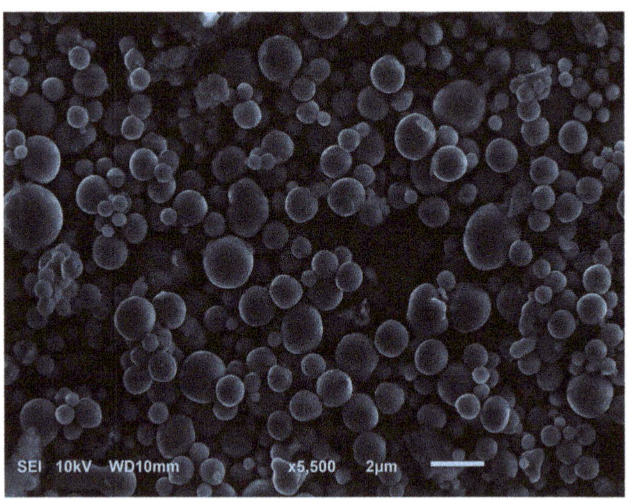

Figure A.1. Scanning electron microscopy image of Ge beads obtained by the laser ablation technique. The particles are characterized by their well-defined spherical shape and with a smooth and uniform surface. Adapted with permission from [2], copyright 2019 by the American Physical Society.

results also reveal that the particles present a well-defined spherical shape with smooth and homogeneous surface. The spectroscopy results shown in figure A.2 demonstrate that the synthesized Ge beads maintained their chemical structure after the laser ablation process. Additional characterizations by energy dispersive x-ray (EDX) analysis have also demonstrated that the synthesized particles have no impurities in their composition, which can be seen in figure A.3. The EDX spectrum of Ge beads shows the presence of only Ge, carbon (C) and oxygen (O). The peak of the carbon in the EDX spectrum is due to the presence of C tape that served as support on which the samples are prepared for analysis. On the other hand, the O peak is very common in EDX spectra and may be related to small traces of oxidation on the surface of the particles or adsorbed O. Therefore, even with careful sample preparation, it is difficult to completely eliminate O from the surface of the samples. Si particles underwent the same characterization procedures, confirming their spherical shape and purity [3]. These results demonstrate that LASLS is an effective technique for synthesizing spherical semiconductor particles suitable for optical tweezers experiments, offering low production costs and a simplified synthesis process.

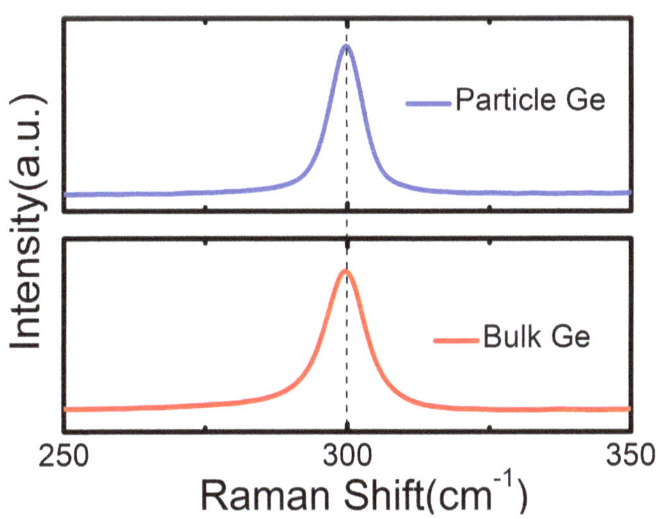

Figure A.2. Representative Raman spectra acquired from the Ge bulk (at the bottom, in red) and a Ge microparticle (at the top, in blue), highlighting the fundamental unstrained Ge Raman line at ~300 cm^{-3}, which corresponds to the bulk Ge phonon mode. Raman spectroscopic analysis revealed that the particles remained chemically unaltered throughout the synthesis procedure. Adapted from reference [2], copyright 2019 by the American Physical Society.

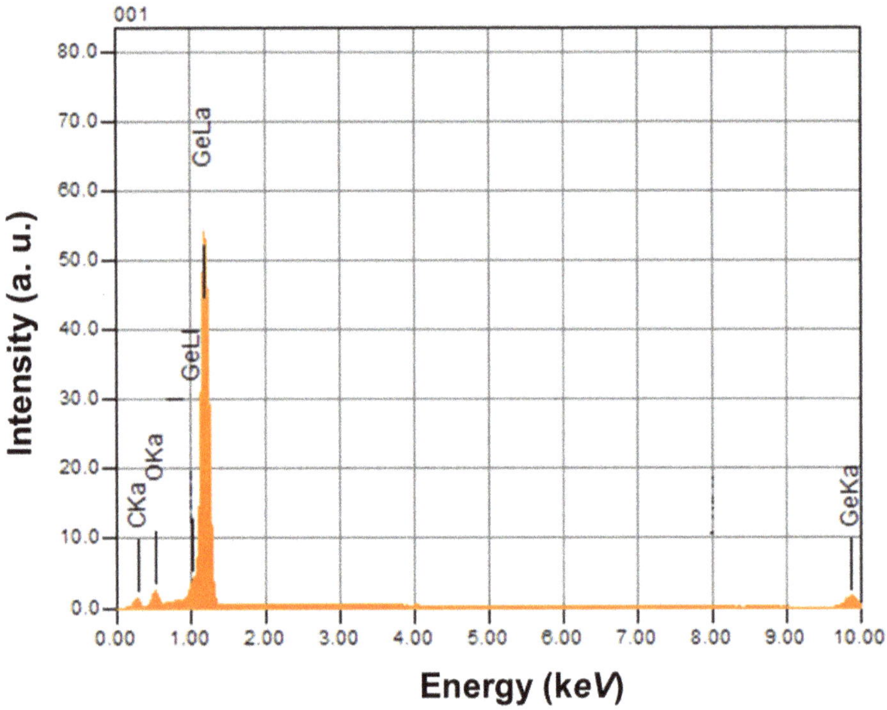

Figure A.3. EDX spectrum from an arbitrary Ge microparticle and showing the presence of only Ge, C and O.

References

[1] Amendola V and Meneghetti M 2013 What controls the composition and the structure of nanomaterials generated by laser ablation in liquid solution? *Phys. Chem. Chem. Phys.* **15** 3027

[2] Campos W H, Moura T A, Marques O J B J, Fonseca J M, Moura-Melo W A, Rocha M S and Mendes J B S 2019 Germanium microparticles as optically induced oscillators in optical tweezers *Phys. Rev. Res.* **1** 033119

[3] Moura T A, Andrade U M S, Mendes J B S and Rocha M S 2020 Silicon microparticles as handles for optical tweezers experiments *Opt. Lett.* **45** 1055

www.ingramcontent.com/pod-product-compliance
Ingram Content Group UK Ltd.
Pitfield, Milton Keynes, MK11 3LW, UK
UKHW051853100926
8556IPUK00005B/597

9 780750 360722